A DISAGGREGATE TRAVEL DEMAND MODEL

A Disaggregate Travel Demand Model

MARTIN G. RICHARDS
MOSHE E. BEN-AKIVA

A Buro Goudappel en Coffeng and Cambridge Systematics Inc. Research Study

SAXON HOUSE | LEXINGTON BOOKS

Published by

SAXON HOUSE, D. C. Heath Ltd.
Westmead, Farnborough, Hants., England.

Jointly with

LEXINGTON BOOKS, D. C. Heath & Co.
Lexington, Mass. USA.

ISBN 0 347 01088 1
Library of Congress Catalog Card Number 74-34528

Printed in Great Britain by Robert MacLehose and Company Limited
Printers to the University of Glasgow

Contents

List of tables

List of figures

Foreword

This book is a slightly modified version of a report entitled *Disaggregate and Simultaneous Travel Demand Models: a Dutch Case Study,* prepared by Buro Goudappel en Coffeng BV, in association with Cambridge Systematics Inc., under contract to the Projectbureau Integrale Verkeers- en Vervoerstudies. The Projectbureau is an organisation established by the Netherlands Government to foster the development of comprehensive transportation planning. It is manned by staff seconded from Dutch Railways (NS), the Water Control and Public Works Department (de Rijkswaterstaat) and the Transport Directorate (het Directoraat-generaal van het verkeer) of the Ministry of Transport and Public Works (Ministerie van Verkeer en Waterstaat), and the Government Physical Planning Agency (de Rijksplanologische Dienst).

The study was based upon data obtained in a large home-interview survey (VODAR) conducted by Buro Goudappel en Coffeng in 1970 under contract to the Rijkswaterstaat.

In order to make the fullest possible use of experience already obtained with direct demand and disaggregate models, Buro Goudappel en Coffeng reached an agreement with Professor Marvin Manheim of the Massachusetts Institute of Technology and Cambridge Systematics Inc., under which Cambridge Systematics undertook to assist in the application of these methods.

The project was conducted under the management of Martin G. Richards of Buro Goudappel en Coffeng with the assistance of Dr Moshe Ben-Akiva of Cambridge Systematics. Other members of Buro Goudappel en Coffeng's staff who played a major role in the study were Meindert Bovens, Koos Mars, Gerard van der Sterre and Pieter Terpstra. Many other members of the company staff gave assistance to the study team, both by providing information as well as by partaking in discussions about the philosophies adopted by the study team and the results obtained with the various model specifications applied. Jan Jetten of the Dienst Verkeerskunde of the Rijkswaterstaat was also an active member of the study team. William Jessiman of Cambridge Systematics provided much guidance throughout the study, as did other Cambridge Systematics staff members. Wally Schouwink and Willy Zuurveld typed the manuscript and the drafts which preceded it.

The guidance of the study by the Projectbureau was the responsibility of Drs C.J. Steilberg and Ir. K. Broersma, without whom the study would never have proceeded and who both provided extremely valuable advice throughout its duration.

Both the Agglomeratie Eindhoven and the gemeente Eindhoven provided much of the data necessary to the study, as did Philips, DAF, Dutch Railways, Zuid-Ooster Autodiensten and the Brabantsche Buurtspoorwegen en Autobusdiensten. The willing assistance of all these organisations is gratefully acknowledged.

Although disaggregate models offer great promise for the future, the development of a fully operational system of disaggregate modelling is still in its infancy; much more research and development is required before such techniques obtain the same degree of usage as traditional aggregate modelling. This is also true of simultaneous travel demand modelling in relation to the traditional sequential system. Whilst disaggregate models and simultaneous models could be viewed as independent concepts, the use of disaggregate techniques can greatly benefit the development and application of simultaneous models, and their union thus has considerable appeal in modelling urban travel behaviour. It is to be hoped that the work described in this book will be of some value to those active in this field of research and will help to stimulate the interests of others in what promises to be one of the most significant advances in travel demand modelling techniques in many years.

Since commencing the study described here, Cambridge Systematics Inc. have undertaken a research contract for the US Department of Transportation using similar techniques, and further research is being carried out at MIT. Buro Goudappel en Coffeng have commenced a major urban transportation study in which the modelling work makes extensive use of the general philosophies described in this book. A second study for the Projectbureau is scheduled to be completed by the end of 1975. This new study is directed towards three specific problems: transferability between geographic areas and also across socio-economic groups; empirical aspects of the transformation of a disaggregate model to an aggregate model; and the application of a simultaneous destination-choice and mode-choice model to a purpose other than work or shopping.

Real success, however, cannot be achieved by individual work in isolation; open discussion, and a free availability of the results of research, in progress and completed, is essential. It was thus very gratifying that the Projectbureau agreed to the publication of the original report in English rather than Dutch.

The permission of the Projectbureau Integrale Verkeers- en

Vervoerstudies to publish their report in this form is gratefully acknowledged. The opinions expressed are, however, those of the authors and are not necessarily those of the Projectbureau. Any errors are, of course, the responsibility of the authors.

Chapter 3 is based extensively on Moshe Ben-Akiva's Ph.D thesis entitled 'The Structure of Passenger Travel Demand Models' and submitted to the Massachusetts Institute of Technology (MIT) in 1973.

The work described in this book was carried out in 1973 and in the first few months of 1974.

Martin G. Richards
Buro Goudappel en Coffeng BV
Postbus 161
Deventer
Netherlands

Moshe Ben-Akiva,
Cambridge Systematics Inc.
238 Main Street
Cambridge
Massachusetts 02142
USA

1 Introduction

1.1 Background

Techniques for the prediction of travel demand are necessary for sound decision-making at each of the different levels of transportation planning, from the development of broad strategies to the detailed design of individual projects. The procedures currently in general use, which have been developed incrementally over the past 20 years and have been widely applied around the world, are, however, being subjected to increasing criticism for their failure to properly address the range of policy issues currently relevant. These failures have been recognised by politicians and technicians alike. An all-party committee of the British House of Commons, after listening to evidence from a number of experts, concluded:

> Among the more serious transportation study weaknesses, in view of the emphasis placed upon traffic restraint in overall policy, is the inability to predict how many people will be deterred from travelling under certain restrictions. In these circumstances we can only underline the former Secretary of State's view that in skilled transportation techniques there must be much more objective assessment of the relationship between traffic flow and parking facilities and the price mechanism on parking. Equally serious, again because of its importance in overall policy, is the inadequacy in predicting how many people might make use of new or improved public transport facilities. We recommend that these possible deficiencies should be rectified by making them a priority area for further research. (House of Commons, 1972.)

Bouchard (1973) reporting to a conference of experts called by the US Highway Research Board (the Williamsburg Conference) identified six shortcomings of current techniques. These were:

1 The models are too time-consuming and too expensive to operate.
2 The models fail, in many ways, to examine all relevant points in the decision-making process.

3 Too much thought is given to the models and too little to the things which are really important in the selection of planning strategies.
4 The models are geared too much to the 1990, or 20-year, situation, when in fact transportation problems are now and projects are now.
5 The technicians themselves do not always understand the models.
6 The models are too data-hungry.

All these criticisms are undoubtedly just; if transportation planning studies, and the models utilised in them, are to fulfil their proper role, then the models must be responsive to current policy issues. They must be valid. They must be timely, i.e. they must provide the information at or before the time when it is needed for the planning or decision-making process. A set of results, no matter how reliable or detailed, is useless if the results become available only after the relevant decision has been taken. Indeed it can be argued that they can have a negative value under such circumstances, since the work involved in their preparation might represent a misuse of available resources. The models must be economical to develop and apply. In fact economical is a relative value in this context, relative to the value of the decisions which will be influenced by the predictions made by the models, and relative to the accuracy of those predictions. Ideally it should be possible to quantitatively specify the accuracy of the predictions, but the accuracy of a future quantity is extremely difficult to determine, especially in travel demand forecasting, which is highly dependent upon the accuracy of the variables utilised in the model and predicted externally – for example, population forecasts by socio-economic characteristics.

Given that current transportation modelling procedures do not satisfy the real demands of the community, and given that these demands have been properly identified, is it possible to devise a set of procedures which more closely approach these demands? The findings of the Williamsburg Conference were quite definitely positive on this point. The conference report states that 'significant improvements in travel forecasting capabilities can be achieved within a period of three years through the use of new techniques based on results of recent research' (HRB, 1973). Another conference finding was that 'substantial improvements in forecasting capabilities can be achieved in the future'.

The conference report goes on to cite disaggregate, behavioural demand models as a direction of great potential. Much of the recent research described at the conference has been directed towards the development of such models designed to be much more behavioural than the traditional models, many of which can be regarded as simply simulation models in the sense that, while reproducing a known situation, they have few, if any, explanatory powers. These are usually correlative models rather than causal (de Neufville and Stafford, 1971).

If a model is to be truly behavioural then it would seem reasonable that it should represent the individual unit of behaviour which, in the context of travel demand, is the household or person. Traditional transportation models have been concerned with groups of persons aggregated into geographic units and can therefore be expected to be primarily correlative models. Despite work as early as 1962 by Oi and Shuldiner demonstrating the deficiencies of such procedures, surprisingly little work has been done in overcoming these deficiencies in operational studies.

The traditional methods have also treated the total demand for travel in a naive manner by usually assuming that it can be broken down into a series of independent elements: trip-end estimation, trip distribution, modal split, and assignment. In fact, since travel is usually a derived demand, it can be expected that the decision-making process over all aspects of a trip is generally a considerably more complex procedure, one which cannot be adequately modelled through the use of simple sequential procedures.

Thus the search for better models has led to two main developments. One is the use of disaggregate data, and the analysis of that data at the level of the behavioural unit, the household or the person. The second is towards a procedure in which the demand for travel can be represented in a single, simultaneous, model.

1.2 The study objectives

The objective of the study on which this book is based was to further the development of behavioural urban travel models, and in particular to implement the methods of disaggregate and simultaneous probabilistic travel demand models in a Dutch context. Despite the apparent geographic limitation, both the process of model development and the study results are relevant on a very much wider scale.

Since this project represents an initial effort in this area, the empirical work is concentrated on two trip purposes only: work and shop. For the work trips, the relevant travel decision for the short to medium term is assumed to be the choice of mode, since it would seem reasonable to assume that both the decision to make the trip and the destination are given. However, for shopping travel we are interested in modelling the choices of frequency, destination and mode simultaneously. The products of this study are a work mode-choice model and a simultaneous shopping destination- and mode-choice model.

The data available to this study are almost unique in so far as urban transportation data are concerned, since each origin and destination is coded to a 10-metre (m) co-ordinate reference. This means that there is no effective geographic abstraction and, consequently, that many of the levels of service data can be derived for each individual observation.

1.3 The structure of the book

Alternative approaches to travel demand modelling are discussed in Chapter 2, and in Chapter 3 the underlying theory and methods of the modelling approach utilised in the study are presented. A general description of the area to which the data utilised in this study relate is given in Chapter 4, as well as a description of those data and of the basic study design. In Chapters 5 and 6 the models estimated for work and shopping trips are presented and evaluated. Chapter 7 contains an evaluation of the project in terms of the models estimated and the data used. Finally a set of conclusions and recommendations on further development and application of travel demand models in the Netherlands is presented in Chapter 8.

2 Approaches to Travel Demand Modelling

2.1 Introduction

It was stated in Chapter 1 that disaggregate and simultaneous modelling procedures are two developments which could overcome much of the current dissatisfaction with conventional travel demand models. In this chapter we endeavour to answer the two questions 'why disaggregate models?' and 'why simultaneous models?'.

The use of the terms 'aggregate' and 'disaggregate' as applied to travel demand modelling is not always totally consistent, and a certain amount of unjustified mystique has been created around them. An aggregate model is a model estimated with a dependent variable which represents a group of observations, whilst a disaggregate model is a model estimated with a dependent variable which represents an observation of a single occurrence (i.e., in travel demand models, a trip). From these definitions we can derive the terms aggregate and disaggregate data. Aggregate data, in travel demand terminology, are generally data grouped to traffic zones, although other forms of aggregation are possible; that utilised in category analysis (Wootton and Pick, 1967) can be regarded as an example of aggregation over household types. Disaggregate data are the basic, individual observations, even though some of the information included within an observation may be at some level of abstraction; in other words it might be based upon some classification system whereby the absolute value of some characteristic is replaced by a group average value (e.g. income classes). Although the dependent variable in a disaggregate model must be an individual observation, some of the independent variables can be represented by aggregate data.

2.2 Why disaggregate models?

For planning purposes, we are concerned with the prediction of the behaviour of aggregates of people. In order to predict a market equilibrium we need a demand function that represents the aggregate demand of all consumers in the market. However, an aggregate demand function is simply the sum of the demand functions of the individual consumers. Demand functions may be developed for different levels of aggregations. The most disaggregate level, or the micro level, is that of the individual consumer, or the single behavioural unit. This is the level at which the behavioural theory on which demand models are based is postulated. Aggregation is achieved by creating groupings or collections of individual consumers. These groupings are often based on geographic location and/or socio-economic characteristics. Obviously, it is not feasible to compute the demand for each individual consumer and accumulate them in order to obtain an aggregate estimate of demand. Hence, for forecasting, some level of aggregation is necessary. The appropriate level is determined from considerations of the purpose of the forecasts, the required detail and accuracy of the analysis, as well as practical and economic factors. In principle, aggregation for prediction purposes can always be performed after estimation provided that detailed, or disaggregate, observations are available; models estimated at the micro level can be aggregated to any level that may be required for forecasting purposes.

In most transportation planning studies we are concerned with predicting the demand for transportation – the demand for travel between pairs of points, origins and destinations. With current network loading or assignment technology we have to represent demand with a relatively high degree of geographic abstraction; thus the origins and destinations have to be grouped so that their true continuous distribution throughout the study area is reduced to a discrete set of points, or zone centroids. Some measure of geographic aggregation is therefore currently necessary, for purely practical and economic reasons, in the network assignment phases of the travel demand modelling process. Traditionally, the same level of geographic abstraction has been maintained throughout the travel demand model system, but there is ample evidence that this is neither desirable nor necessary during model estimation.

In urban transportation planning studies, data are conventionally collected at the disaggregate level, usually by means of a home-interview survey, conducted with a sample of households drawn with a uniform sampling rate throughout the study area. The data are then aggregated geographically using a traffic-zone system, which is defined largely on the basis of accessibility to various transport facilities and expanded to represent the total population. The travel demand models are then estimated using the aggregate, or averaged, data, with the traffic zone as the observation unit. Transport level of service data relating to a traffic zone are measured relative to a single point, or centroid. This means that the population, or consumers, within a zone are assumed to be located at one point in space. This is equivalent to the use of the mean of the category in an aggregation by socio-economic categories. Aggregation before the model construction phase of the analysis will cloud the underlying behavioural relationships and will result in a loss of information. An aggregate model which is based on averages of observations of socio-economic types and geographic location does not necessarily represent an individual consumer's behaviour, nor the average behaviour of the group under a variety of conditions. Thus there is no reason to expect that the same relationship would hold in another instance or another location. Models estimated directly from individual observations, without aggregation, represent the typical behaviour of individual consumers. To the extent that consistent consumer behaviour for similar circumstances can be assumed, a model estimated with data on a set of consumers in one area can be used to predict behaviour of other consumers in another area, or at a different point in time.

Estimation and prediction results of aggregate models have to be very carefully interpreted due to the risk of 'ecological fallacy' (de Neufville and Stafford, 1971) which arises because correlations among aggregate variables do not necessarily reflect actual behavioural association at the level of the individual behavioural units. Robinson (1950) showed that ecological correlations should not be utilised to explain individual behaviour.

A further problem which, whilst being generally relevant, is particularly important in aggregate models is that of the functional form of those independent variables the value of which is dependent upon the aggregation scheme used. If the relationship between the dependent variable and such independent variables is non-linear, then the functional form and the coefficients are likely to vary with different aggregation systems. Thus an aggregate model which is non-linear in such variables is valid only for the system of aggregation for which it was estimated. Yet Stowers and Kanwit (1966) have argued that the assumption that all relationships are linear can be incorrect.

Oi and Shuldiner (1962) demonstrated some of the major problems associated with geographic aggregation in modelling the number of trips produced at the home. Geographic, or zonal, aggregation before estimation implies an averaging over the individual behavioural units which, in the context of travel demand, are either households or persons within a zone. This averaging, however, is not a problem if the households (or persons) within each zone exhibit a high degree of homogeneity in those characteristics relevant to travel demand and which are thus incorporated as explanatory variables in the models. Such homogeneity is necessary if the different characteristics of all the individual behavioural units within a zone are to be properly represented by the zonal mean value for each individual characteristic, be they socio-economic attributes or level of service (transportation systems) attributes.

The conventional trip production models based on zonal aggregations explain only differences between zones. Perfect homogeneity is extremely difficult, if not impossible, to achieve in practice, and Fleet and Robertson (1968) have demonstrated that the variability within zones can be very much greater than the variability between zones. Thus, as stated by McCarthy (1969), zonal aggregation obscures the true relationship between household trip production and household socio-economic characteristics, making it difficult to develop general trip production models applicable to more than one study area. If the trip production models are not transferable from one area to another, then there must also be serious doubt about their validity when applied to different zoning systems within the same study area, as implicitly (or explicitly) occurs in an area subject to extensive change between the base, or survey, year and the plan year.

It should be noted that in this discussion we are assuming that any demand model – aggregate or disaggregate – is estimated for a group of travellers possessing homogeneous behaviour. This means that both the functional form and coefficients of the model are the same, although the explanatory variables do not necessarily have equal values. Equal values of the variables is what was meant by homogeneity in the previous paragraph.

Another aspect of aggregate modelling is that, given that there is considerable variability within zones, much valuable information is lost through the averaging implied in the aggregation process. The discussion above has been concerned with trip production models, but similar arguments can be applied to other phases of the traditional travel demand modelling system.

Consider, for example, the distance of a bus stop from a residential address. This distance can be expected to have a major effect on the selection or rejection by a person of bus as the mode of transportation for a particular trip. It can be expected that in a given zone some people will live very close to a bus stop, while others will live some distance from one. Within the one zone there will thus be a high degree of variability in the walking distance between home and the nearest bus stop, but in a zonal model usually only the average distance is utilised. If the average value were similar for all zones, as could well occur – i.e. the variability between zones may be very much less than within zones – then the distance between the residential address and the nearest bus stop may be found to have little or no effect on mode-choice. This variable would thus, erroneously, be excluded from the ultimate model specification. This example indicates that a disaggregate model could be more responsive to policy issues than an aggregate model.

An example of the aggregation problem in traditional trip-distribution models is that of predicting the volume of trips remaining within a zone, i.e. having both origin and destination in the one zone. In many transportation planning studies the intra-zonal flows are relatively large, yet they must be estimated by assuming that all trips within a zone have an identical length, or price, whereas in fact they can vary from very short trips to relatively long trips, depending on the coarseness of the zoning system. The relevance of this argument is frequently questioned and in some studies is rejected and only inter-zonal flows considered, but this approach hardly solves the problem. Certainly it is not behavioural.

Thus travel demand models using data aggregated in any way, other than to purely homogeneous groups (homogeneous, that is, with regard to any of the explanatory variables included in the model), can be faulted on both technical and economic grounds. They are unlikely fully to explain the behaviour of the individual travellers, and thus they are likely to be poor prediction models. Furthermore, they are highly inefficient in their use of data. These major deficiencies can be avoided through the development of models based on the individual behavioural unit, the household or the person. Not only does this enable us to understand, and thereby to model, those factors which really influence the travel behaviour of the individual units, but it also enables us to make more efficient use of the data available. This latter point is highly attractive because in modelling travel demand we are endeavouring to explain differences in observed behaviour. The greater the variability in the observed data, the greater are the chances the model builder has of explaining the observed differences between the individuals, and thus the greater the confidence we can have in his models. By considering the individual unit of travel behaviour, by dealing with totally disaggregate data, the model builder is assured of a much greater degree of variability than when he has to confine himself to aggregated data sets. Through more efficient use of the information available, it should prove possible to reduce the volume of observations required for model construction, and thus to reduce the costs of transportation planning studies. Hillegas

(1969) and Wickstrom (1971) have both noted that data collection and processing usually consume as much as half of the total study budget.

By considering individual behaviour and seeking out true causal relationships, we have a much greater chance of developing a general model which can be transferred from one area to another, or from one population to another. The fact that there has been very little success to date in developing a general travel demand model which can be utilised, without changing the values of any of the coefficients, in different areas can probably be attributed to the fact that models have been aggregate and have therefore not been truly behavioural.

Warner (1962) was one of the first to use disaggregate data in mode-choice modelling when he developed a probabilistic, binary mode-choice model. Since then experience has been gained in disaggregate modelling of mode-choice and other travel choices in a number of studies (e.g. de Donnea, 1971; Reichman and Stopher, 1971; CRA, 1972; PMM, 1972; Ben-Akiva, 1973; Watson, 1973), which all suggest that disaggregate travel-demand modelling is a feasible modelling approach. However, while a model should desirably be estimated at the disaggregate level, it is not always possible to utilise the same model directly for the preparation of predictions, since forecasts of data on the explanatory variables are not usually available at a disaggregate level. Thus a model estimated as a disaggregate model normally has to be applied as an aggregate model for prediction work. This transformation is not without its problems, but in fact such problems are equally applicable to an aggregate model, a fact which is normally overlooked. The application of a model estimated using disaggregate data for aggregate forecasting is in principle a very simple procedure. The practical problem involved is the prediction of the distribution of the independent variables and not simply their means.

The problems associated with the transformation of a disaggregate model into an aggregate model are discussed in section 3.6.

2.3 Why simultaneous models?

In most urban transportation situations, the traveller has a choice of trip frequency, time of day, destination, mode, and route. In a simultaneous model the traveller is presumed to choose all of these elements (make all these component decisions) simultaneously in a single 'joint' choice; thus all attributes of the entire trip, and of all alternative combinations of trip possibilities, are considered simultaneously. In a recursive, or sequential, model the traveller is presumed to make his decision in a sequence of steps. A recursive model structure is defined here as one in which there are no feedback loops or mutual interactions; thus they have a hierarchical structure (de Neufville and Stafford, 1971). For example, in the conventional Urban Transportation Model System (UTMS), the implicit assumption is that the first decision is whether (and how often) to make a trip – i.e. trip frequency – then, conditional on the assumed frequency, which destination to choose, followed by the mode with which to make the trip, and, finally, the route. Thus the conventional travel demand models represent not only the behavioural assumption that a traveller's decision is sequential, and not simultaneous, but also that it follows one particular sequence.

A trip for a specific purpose is characterised by its origin *(i)*, destination *(d)*, and mode of travel *(m)*. We are normally interested in predicting the volume of trips, V_{idm}. The demand model of the conventional UTMS is usually structured into the following sequence of steps:

- the trip generation model predicts the total volume V_i from an origin i regardless of destination or mode:
- the trip distribution model distributes the volume V_i among the destinations to predict V_{id}, frequently regardless of mode;
- finally, the modal split model predicts V_{idm}.

From the point of view of the individual trip-maker, the above sequence represents a conditional decision-making process. For example, the choice of mode is conditional on the choice of destination, but the destination choice was made without consideration of chosen mode.

From a theoretical point of view a conditional decision-making process is not completely realistic for some trip purposes. It has been argued that the trip decision should be modelled simultaneously without resorting to an artificial decomposition into sequential stages (Kraft and Wohl, 1967). For example, in the case of a shopping trip it is, in general, more realistic to assume that the decision to make a trip, the choice of destination, and the choice of mode are interdependent and therefore should be determined simultaneously. A simultaneous decision-making process is represented by a single model that predicts V_{idm} directly.

Attempts to develop simultaneous models have been made using the conventional approach of aggregate demand analysis, where the quantity demanded is taken as a continuous variable (e.g., Kraft, 1963; Quandt and Baumol, 1968; Domencich *et al.*, 1968; Plourde, 1968). All the disaggregate models developed could be used either for a single stage of the UTMS (e.g., Reichman and Stopher, 1971), or, more recently, for all the stages, but again assuming a recursive structure (CRA, 1972).

Recently, disaggregate simultaneous models have been developed by Ben-Akiva (1973) and Cambridge Systematics (1973). This work clearly demonstrates the feasibility and desirability of disaggregate simultaneous travel demand models. The study described in Chapters 4, 5 and 6 takes this development a stage further.

Although from a behavioural point of view it would seem best to assume a simultaneous choice model for all those situations where there are no behavioural arguments for any particular sequential model, there may be practical problems. Because of the large number of alternatives confronting a traveller and the large number of attributes which describe an alternative trip, a simultaneous model can become very complex. This raises some important issues concerned with the practicality of a simultaneous model, and the sensitivity of travel predictions to the simplifying assumption of a recursive structure. However, these problems can be resolved other than by assuming a sequential model, in any of several ways, including reducing the set of alternative choices considered by a *priori* or empirical arguments, and reducing the number of attributes a traveller is assumed to consider.

In addition to the conceptual appeal of simultaneous models, the use of such models could also result in lower computing costs, since the time-consuming iterations customarily done within a distribution model to achieve equilibrium between productions and attractions could be eliminated.

2.4 Summary of the alternative approaches to travel demand modelling

The previous sections can be summarised by noting that there are four basic alternative approaches to travel demand models:

(1) aggregate simultaneous models – for example, Charles River Associates' San Francisco demand model (CRA, 1967);
(2) aggregate recursive models – for example, the traditional UTMS models;
(3) disaggregate simultaneous models – as estimated by Ben-Akiva (1973), for example;
(4) disaggregate recursive models – for example, the disaggregate modal split models (Reichman and Stopher, 1971; PMM, 1972), and the more recent demand model estimated by CRA (1973).

In addition to these four alternative modelling or model estimation approaches, there are also two basic ways to use any of these estimated models in predictions. The two basic alternative ways are:

(1) direct – directly predicting the volume of trips V_{idm}, for example (this is the predictive counterpart of simultaneous model estimation);
(2) indirect – predicting V_i first, then V_{id}, and finally V_{idm}, for example (this is the forecasting counterpart of recursive model estimation).

The current trend in travel forecasting and the approach pursued in this study is toward disaggregate simultaneous models and direct forecasting.

3 Methodology

3.1 The choices in, and the demand for, travel

The complexity of travel demand is apparent from the way we characterise a trip, its origin, its destination, the time of day when it is made, the mode of travel, the route, and the trip purpose.

Trip ends themselves are spatially disaggregated; a trip is made from a specific origin to a specific destination. For a work trip the origin and the destination are fixed by the choices of the residential location of the household and the trip-maker's place of work. Whilst the one end of most other trips is fixed by the location of the trip-maker's home, a choice usually exists between several alternative destinations. For example, a shopping trip could be made either to a local grocer or to a more distant shopping centre.

The point in time during the day when a trip is made can be influenced by the effect of peak periods and the congestion usually associated with travel during a peak. Whilst starting and finishing times are not totally within the control of the majority of workers, they can still decide to arrive early and leave late in order to avoid travelling during the peak period, and some degree of freedom is provided by employers operating flexible working-time systems (flexitime). Optional activities are less constrained, and the traveller can normally exercise a reasonable degree of choice as to when he makes a trip to and from such activities.

In most urban areas in the Netherlands a traveller has a choice between public transport, private car, bicycle, moped and walking. A traveller can also sometimes choose between driving his own car, being driven, or participating in a car pool; some travellers can also choose to make a single trip utilising a sequence of different modes. A typical example of such a complex trip is the commuter who drives his car to a suburban railway station, parks the car, takes the train to the city centre and finally travels by bus or tram to his place of work. In dense networks, the traveller can usually choose between several alternative routes, especially when travelling by car, bicycle or moped. Speed, congestion, distance, safety, driving comfort and reliability can all affect the choice of route.

Thus for a given trip the traveller is faced with a set of choices: destination, time of day, mode, and route. In addition, a potential trip-maker has to decide whether or not to make a trip, or how often to make a trip for a given trip purpose. For example, families can shop for groceries with different frequencies – every day, twice a week, once a week, etc. – and the shopping trip can be undertaken by one or more members of the family.

A trip is rarely, if ever, made for its own sake, but for some purpose which can be served at its destination, in order to gain some benefit, or utility. Thus the classification of trips by their purpose does not represent an inherent characteristic of the trip itself, but rather stems from the recognition that the demand for trips is a derived demand, as pointed out in Oi and Shuldiner (1962) and Kraft and Wohl (1967). Therefore, trip-makers' behaviour need not be the same for different trip purposes. Given the normal practice of defining classes of trip purpose such as work, firm's business, recreation, school, social, shopping, and personal business, it is reasonable to hypothesise that there is negligible substitution between trips for different purposes unless the corresponding activities undertaken at the trip destination can themselves be considered as substitutes for each other.

It is usual practice in urban transportation planning studies (UTPS) to link trips to remove secondary purposes, such as 'change mode', so as to give a single, one-way trip by the dominant mode from the true origin to the ultimate destination (BPR, 1965). However, given that the demand for travel is a derived demand, this definition is not fully appropriate. A trip from home to an activity is usually accompanied by a trip from that activity back home, forming a simple tour, or chain, of two trips; under more complex circumstances, more than one activity can be undertaken, or satisfied, during the course of a complex tour composed of three or more trips. In a complex tour the trips to and from each destination are generally interdependent, since decisions made with regard to the first leg both constrain and influence the decisions made with regard to the remaining legs, and the alternatives available for those remaining legs can be expected to influence the decisions made concerning the first leg. For example, someone who goes from his home to shop by car is highly likely also to return home by car. If, on this tour, he chooses to make another stop, for shopping or any other activity, his choice of the two destinations cannot be considered as being independent of each other, nor can his choice of mode: he has probably considered all elements of his travel decision prior to starting out on his trip.

If we consider a tour as the basic unit, instead of the more conventional unit, the trip, the complexity of travel demand modelling increases further. A single trip is described by its origin, destination, time of day, mode, route, and purpose. A tour has a single base point, or origin, but can have a mixture of several different destinations, times of day, modes, routes, and purposes.

Simple tours, e.g., home–work–home, home–shop–home, etc., have in most cases two identical legs and therefore pose no additional problems for modelling. In the few cases in which the mode of travel for the two legs is not the same – e.g. going shopping by bus and returning by taxi – the relatively few possible combinations of modes could be handled by defining them as separate alternatives. However, for some complex tours the dimensions of choice – that is, the number of alternative trip options – can be extremely large. Fortunately, the majority of urban trips are elements of a simple two-trip tour.

3.2 Choice theory

In general, models describing consumer behaviour are based on the principle of utility maximisation, subject to resource constraints. That is, they are based on the assumption that the consumer will endeavour to maximise the benefits he can obtain within the limitations of the available resources – usually time and money. The demand approach of economic consumer theory assumes a selection of quantities from a set of commodities, and the quantities demanded are treated as continuous variables. Choice theories consider a selection from a finite set of mutually exclusive and exhaustive alternatives.

In the demand approach, with a continuum of alternatives, a deterministic behaviour is assumed, except for a random disturbance added in estimation to explain variations among people and to account for factors not incorporated. In choice theory, with qualitative or discrete alternatives, a probabilistic behaviour is assumed which explains observations of different choices for the same set of observed independent variables.

The travel demand of a single consumer is more appropriately viewed in a choice context rather than in the traditional demand analysis framework. Since travel choice can better be described as a probabilistic choice from a discrete set of mutually exclusive and exhaustive alternatives than as the selection of a quantity of a commodity, the travel demand models developed in this study rely on probabilistic choice theories, which are reviewed in Ben-Akiva (1973), Chapter 3. See also Luce and Suppes (1965) and CRA (1972).

The consumer is visualised as selecting that alternative which maximises his utility. The probabilistic behaviour mechanism is a result of the assumption that the utilities of the alternatives are not certain, but are random variables determined by a specific distribution.

If we denote the utility of alternative i to consumer t as u_{it}, the choice probability of alternative i is therefore:

$$P(i{:}A_t) = \text{Prob}\left[u_{it} \geqslant u_{it},\ \forall j \epsilon A_t\right] \tag{3.1}$$

where A_t is the set of alternative choices available to consumer t; this set of alternatives is both mutually exclusive and exhaustive such that one, and only one, alternative is chosen. This expression basically means that the probability of choosing one alternative out of the complete set of possible alternatives, $P(i{:}A_t)$, equals the probability of the utility of that alternative being greater or equal to the utility of any of the other alternatives in the full range of alternatives. The deterministic equivalent of this theory is simply a comparison of all alternatives available, and the selection of that alternative having the greatest utility.

The utilities are essentially indirect utility functions which are defined in theory as the maximum level of utility for given prices and income. In other words, the utility u_{it} is a function of the variables which characterise alternative i, denoted as X_i, and of the socio-economic variables describing the consumer t, denoted as S_t. Assuming that the random components of the utilities can be expressed as additive disturbances, we can write:

$$u_{it} = u_i\,(X_i,S_t) + \mu_{it} \tag{3.2}$$

where μ_{it} is an additive random element.

A simple example of such a function is a utility function for a mode-choice model composed only of travel time and car availability. The travel time can be expected to vary between alternatives, whilst car availability remains constant for a given person. The travel time would thus explain the choice of mode for a given person, and the car availability would explain differences in mode choice between persons having an identical set of levels of service for each available mode but different levels of car availability.

Substituting equation (3.2) in equation (3.1) we get:

$$P(i{:}A_t) = \text{Prob}\left[U_i(X_i,S_t) + \mu_{it} \geqslant U_j(X_j,S_t) + \mu_{jt},\ \forall j \epsilon A_t\right] \tag{3.3}$$

Rearranging terms, this can also be written as:

$$P(i{:}A_t) = \text{Prob}\left[\mu_{jt} + \mu_{it} \leqslant U_i(X_i,S_t) - U_j(X_j,S_t),\ \forall j \epsilon A_t\right] \tag{3.4}$$

This expression implies that the mathematical form of the choice model is determined from the assumption about the joint distribution of the random elements.

Assuming that the random elements of the alternatives are *independently* and *identically* distributed with a reciprocal exponential distribution, it can be shown (CRA, 1972; Ben-Akiva, 1973) that the choice model takes the form of the multinomial logit model as follows:

$$P(i:A_t) = \frac{e^{U_i(X_i,S_t)}}{\sum_{j \in A_t} e^{U_j(X_j,S_t)}} \qquad (3.5)$$

If we substitute in the logit model (3.5)

$$V_{it} = e^{U_i(X_i,S_t)} \qquad (3.6)$$

then we get

$$P(i:A_t) = \frac{V_{it}}{\sum_{j \in A_t} V_{jt}} \qquad (3.7)$$

which is the form of model developed by Luce (1959) in which the utility V_{it} is 'strict utility' (Marschak, 1959; Luce and Suppes, 1965).

The assumptions about the distribution of the random elements are assumptions which are of questionable validity in a number of choice situations. These are discussed here in section 3.5.3, and elsewhere by, for example, Luce and Suppes (1965), and Manski (1973). Other assumptions, which might be regarded as being of greater validity, are, however, more complex and cannot currently be implemented in studies such as this due to their demanding computational requirements. For this reason logit is the only choice model which has been extensively used to date for multiple choice situations. A fuller description of logit is given in subsequent sections of this chapter.

The coefficients of each of the variables comprising the utility functions are estimated over a cross-section of consumers using observations of the actual choices made and determination (or observation) of the alternatives available but not chosen. The observed dependent variable thus has a value of zero when an alternative was not selected and of one when an alternative was selected. When applied to forecasts, the model will give the probability of each of the alternatives being selected; the sum of these probabilities must clearly equal 1·0.

3.3 The travel choices

Given that for the individual trip-maker the decision to make a trip for a given trip purpose consists of several choices (trip frequency, destination, time of day, mode of travel, and route), we need to predict the volume of trips V_{idhmr}, from origin i, to destination d, during time period h, using mode m, and via route r. Or, in a probabilistic choice approach, we are interested in predicting the following joint probability:

$$P(f,d,h,m,r{:}FDHMR_t)$$

This is defined as the probability that person t will decide to make a trip with frequency f, to destination d, during time of day h, using mode m, and via route r, from the full set of available alternatives of frequencies (F), destinations (D), times of day (T), modes (M), and routes (R), available to that individual (t), i.e., $FDHMR_t$. In general the utility U_{fdhmrt} of a given trip can be denoted thus:

$$U_{fdhmrt} = U_{fdhmr}(Z_{fdhmrt}) \tag{3.8}$$

where
$$Z_{fdhmrt} = \left[L,E,S\right]_{fdhmrt} \tag{3.9}$$

and L = level of service characteristics such as time, travel cost, safety, comfort, etc.;

E = spatial opportunity characteristics describing the availability and the intensities of the activities for which a trip is made, e.g., employment, land use, etc.;

S = socio-economic characteristics of the trip-maker, such as income, education, sex, etc.

It should be noted that L and E both represent attributes of the alternatives which can be more generally denoted by X, as in (3.2). Thus, for a mode-choice model, in which in the interests of simplicity we consider only mode (m), destination (d), and frequency (f), the utility function can be written:

$$U_{mt} = U_m\left[X_{m|d,f}, S_t\right] \tag{3.10}$$

where $X_{m|d,f}$ means that the values of the variables for mode m are for a given destination and frequency. This form of model is described as a conditional model, in that it is applied to the estimation of mode choice, given that the frequency and destination are predetermined, or conditional upon their being so.

For a simultaneous mode- and destination-choice model the utility function is written as follows:

$$U_{mdt} = U_{md}\left[X_{md|f}, S_t\right] \qquad (3.11)$$

In this case the selection of destination and mode is conditional upon a trip being made, or more precisely on the trip frequency, f.

Clearly the set of alternatives available to any given person varies between choice situations, and thus the set of alternatives considered for any given person will be different for different models. In a mode-choice model, which assumes the destination-choice as given, the set of alternatives is all the possible modes to the given destination. In a destination-choice model the set of alternatives is all the possible destinations for a given trip purpose. In a simultaneous destination- and mode-choice model the set of alternatives will include all the possible combinations of modes and destinations.

The set of alternatives may also differ for otherwise identical persons at different locations. For example, a shopping centre served directly by bus from one neighbourhood could be considered as being a valid alternative for a non-car owning household living in that neighbourhood, whereas it would not be a valid alternative for a similar household living in another neighbourhood lacking a suitable bus link. It is however possible to assume, superficially, that all alternatives are valid and then require the model to predict the fact that a given alternative is not feasible by forecasting a zero, or near zero, probability of choice.

If an alternative has zero or very close to zero choice probability, its inclusion or exclusion from the set of alternatives will have negligible effects on the estimation and forecasting results of the model (see section 3.5.6.). This means that we can attribute to any person any alternatives which are highly unattractive and which will be predicted by the model to have zero or a very small probability of choice. Alternatively, the set of alternatives considered can be restricted to those truly relevant and all others can be deleted with no significant effect on the model. The reduction of the set of alternatives to the relevant set, i.e. those alternatives with non-negligible choice probability, has, however, major practical advantages, since it enables the achievement of major savings in data collection costs as well as in computational costs.

The determination of the relevant alternatives does, however, create some practical problems. One solution is to endeavour to determine the apparent alternatives from the observed trip making pattern. For example, if a taxi is never or rarely used, then it could be rejected as a relevant alternative mode, or if a shopping centre at location j is not visited by travellers from area i in the sample, then location j could be rejected as a relevant destination for persons at i. An alternative is to identify during data collection the possible, but rejected, alternatives in addition to that actually selected.

In general, the set of alternatives used for forecasting does not have to be absolutely identical with the one used in estimation, though some correspondence seems desirable.

3.4 The alternative structures

If we consider the choice of destination and mode, we can denote the complete set of all alternative combinations of modes and destinations as DM. The combination of a specific destination d and a specific mode m is denoted by dm, and the set of modes available to destination d can be denoted M_d. If the modes and destinations available had no common attributes and the two choices were independent, then M_d is independent of d, and could be denoted simply as M. This, however, is not a very realistic assumption, since many of the attributes affecting travel choice are factors in all travel choices, be it the choice of frequency, destination or mode. A typical example of such an attribute is travel time.

For the purposes of illustration of the model theory, let us assume that our prime concern is the prediction of the joint probability of choosing d,m from the complete set of choices DM: $P(d,m:DM)$. Subscript t has been dropped in order to simplify the presentation. If we were to assume that the two choices of mode and destination were mutually independent, we could write the following independent structure (as reasoned in the previous paragraph, this is an unrealistic structure for travel demand, but it is presented for the purpose of comparison with other structures):

$$P(d:D) = \text{Prob}\left[U_d \geqslant U_{d'},\ \forall d' \in D\right] \tag{3.12}$$

$$P(m:M) = \text{Prob}\left[U_m \geqslant U_{m'},\ \forall m' \in M\right] \tag{3.13}$$

$$P(d,m:DM) = P(d:D)\ \ P(m:M) \tag{3.14}$$

where D = the set of alternative destinations;
M = the set of alternative modes;
U_d = the utility from destination d;
U_m = the utility from mode m.

In other words, we could predict the probability of choosing *dm* by considering independently the probability of choosing d from D, and of choosing m from M. The probability of choosing *dm* is then given by multiplying the two independent probabilities.

Alternatively, we could consider a conditional decision-making process in which, for example, the destination is chosen first, and then, conditional on the choice of destination, a mode is chosen. In this case we would write the following recursive structure:

$$P(d{:}D) = \text{Prob}\left[U_d \geqslant U_{d'}, \forall d' \in D\right] \tag{3.15}$$

$$P(m{:}M_d) = \text{Prob}\left[U_{m|d} \geqslant U_{m'|d}, \forall m' \in M_d\right] \tag{3.16}$$

$$P(d,m{:}DM) = P(d{:}D)\, P(m{:}M_d) \tag{3.17}$$

where M_d = the set of alternative modes to destination d
$U_{m|d}$ = the utility from mode m, given that destination d is chosen.

Finally, if we assume that the choice of mode is dependent on the choice of destination and *vice versa,* we would write the following simultaneous structure:

$$P(d{:}D_m) = \text{Prob}\left[U_{d|m} \geqslant U_{d'|m}, \forall d' \in D_m\right] \tag{3.18}$$

$$P(m{:}M_d) = \text{Prob}\left[U_{m|d} \geqslant U_{m'|d}, \forall m' \in M_d\right] \tag{3.19}$$

where D_m = the set of alternative destinations by mode m.

Whilst with the independent and recursive structures the joint probability could be predicted by multiplying the structural probabilities, in a simultaneous structure the two conditional probabilities provide insufficient information to predict the joint probability. Therefore, we need either to estimate a marginal probability (the probability of one event of a joint distribution), say $P(d{:}D)$, or to estimate directly the joint probability. The problem with the first approach is that we need to define a utility U_d, whereas we originally specified $U_{d|m}$. The second approach requires a specification of the joint utility U_{dm} when we consider each combination dm jointly as a single alternative. This approach is more logical, since it corresponds with the notion of a simultaneous choice. Hence, in the simultaneous structure, we need to estimate the following choice probability:

$$P(d,m:DM) = \text{Prob}\left[U_{dm} \geqslant U_{d'm'}, \forall d',m' \in DM\right] \qquad (3.20)$$

Given the joint probability, we can derive any desired marginal or conditional probability. For example:

$$P(m:M) = \sum_{d \in D_m} P(d,m:DM) \qquad (3.21)$$

$$P(d:D_m) = \frac{P(d,m:DM)}{P(m:M)} \qquad (3.22)$$

3.5 The multinomial logit model

3.5.1 *The model*

In the previous section the demand for travel was characterised as a choice from a set of discrete, or qualitative, alternatives, just as the total amount of travel can be regarded as a choice among different trip making frequencies. Since a potential traveller can decide not to make a trip at all, a no-trip alternative must also be included in the set of alternatives confronting him. Consequently, for any given traveller, one, and only one, alternative can be chosen from this set but the set of alternatives can be assumed to be different for different travellers.

We are interested in the probability of each alternative being chosen; thus in the aggregate we are interested in the share of each alternative. Clearly the probability, or the share, of any alternative must lie between zero and one (inclusive) and the sum of the probabilities (shares) of all alternatives must equal unity. The model used should be able to handle any number of alternatives and different sets of alternatives for different people; this is a requirement that is not met by the technique of discriminant analysis which has frequently been used for mode-choice models, e.g., Quarmby (1967); McGillivray (1969); Plourde (1971).

The basic choice model used throughout this study is the multinomial logit model. Other models which might be considered to be superior from a theoretical point of view, such as a multiple probit model, are more complex and it is doubtful whether the added expense of more sophisticated choice models is justified (de Donnea, 1971; Ben-Akiva, 1973; Watson, 1973).

The logit model is written as follows:

$$P(i{:}A_t) = \frac{e^{U_{it}}}{\sum_{j \epsilon A_t} e^{U_{jt}}} \tag{3.23}$$

where t = a behavioural unit (in the case of this study, a person) $= 1, 2, \ldots, T$;

A_t = the set of relevant alternatives for behavioural unit t;

$P(i{:}A_t)$ = the probability that behavioural unit t will choose alternative i out of the set A_t;

U_{it} = the utility of alternative i to behavioural unit t.

The utility U_{it} is a function of the characteristics of alternative i (or generalised 'prices') and the socio-economic characteristics of behavioural unit t (or 'income'). The characteristics of the alternatives could include the travel time to the alternative, the parking charge or congestion at that alternative, or some measure of its attractiveness, such as the number of employees. The socio-economic characteristics include income and other variables, such as family life cycle, education, etc., which can account for differences in tastes, etc. In essence this is thus an indirect utility function which is defined as a function of prices and income. Alternatively, it can be viewed as an expenditure function which is a function of prices. Hence, the function U_{it} can be expressed in the form:

$$U_{it} = U_i(X_i, S_t) \tag{3.24}$$

where X_i = a vector of characteristics of alternative i;

S_t = a vector of socio-economic characteristics of behavioural unit t.

It is assumed that the function U_{it} is linear in the parameters:

$$U_{it} = X_{it}{}'\theta = \sum_{k=1}^{K} X_{itk}\theta_k \tag{3.25}$$

where X_{it} = a $K \times 1$ vector of finite functions which are constructed from the various X_i and S_t variables and are different from one alternative to the other (this definition is explained in detail in the following section)

$= (X_{it_1}, X_{it_2}, \ldots, X_{itK});$

θ = a $K \times 1$ vector of coefficients to be estimated for each model

$= (\theta_1, \theta_2, \ldots, \theta_K).$

This assumption can be justified if we consider the Us to be the combination of equivalent costs for the different alternatives, i.e., that the coefficients reflect the conversion factors from the units of the particular parameter to monetary units. For example, the coefficient of travel time would reflect value of time in guilders per hour, hence converting the time term to equivalent guilders. This assumption is required since available estimation procedures require linearity in the parameters.

Equation (3.23) can now be expressed as follows:

$$P(i{:}A_t) = \frac{e^{X_{it}'\theta}}{\sum_{j \epsilon A_t} e^{X_{jt}'\theta}} \qquad (3.26)$$

3.5.2 *The specification of the variables*

A variable in X_{it}, say X_{itk}, can be specified either as a 'generic variable' or as an 'alternative specific variable'. If the variable X_{itk} appears only in the utility function of alternative *i*, then it has a value of zero for all other alternatives; that is:

$$X_{jtk} = 0, \forall j \neq i \epsilon A_t$$

This variable is thus specific to *i* and is termed an alternative *i* specific variable.

If the variable X_{itk} appears in the utility functions of all the alternatives, it is termed a generic variable. Consider, for example, the variable of travel time in a mode-choice model; if the travel times by the different modes are assumed to have a common coefficient, or weighting, in their respective utility functions, then travel time is specified as a generic variable. The variable X_{itk} will then take the value of travel time by mode *i*. If, on the other hand, we assume the coefficients of travel time to be alternative, or mode, specific, then we introduce a series of mode specific variables. In this case we assume, a *priori*, that the weight of travel time in the respective utility functions for each mode need not be constant across all modes. The mode *m* travel time variable will then be specified as follows:

$$X_{itk} = \begin{cases} \text{travel time by mode } m, & \text{for } i = m \\ 0, & \text{for } i \neq m \end{cases}$$

and if there are M alternative modes, we will have a total of M travel time variables and coefficients.

If all the variables in the model are generic, then the model can be described as an abstract-alternative model (Quandt and Baumol, 1966; CRA, 1972). Since there are no variables present in such a model that relate to any specific alternative, but only variables relating to characteristics common to all the alternatives, an abstract-alternative model can be applied for forecasting purposes to situations substantially different from those used for model estimation. An abstract-alternative model is thus highly suited to the evaluation of systems not yet in use.

Alternative specific variables are relevant only when there is direct correspondence among the alternatives available to different people, e.g. alternative modes. The specification of all the variables in a model as alternative specific implies that the utility functions are alternative specific as well.

When there is little correspondence among the sets of alternatives available to different people, we can only use generic variables. For example, if the set of alternative shopping destinations at one origin are entirely different from the destination set at a different origin, we can say that there is no correspondence among the sets of alternatives and thus that those alternatives can be described only through the use of generic variables. Another example is that of alternative routes, which are inevitably different for different trips.

CRA (1972) explain this by reference to ranked and unranked alternatives. A ranked alternative is one in which each choice within a set of alternatives can be positively identified and is the same for all individuals. Thus trip cost as a variable in a mode-choice model can usually be described as ranked. Denoting trip cost as $OPTC_{jt}$, where subscript j indicates the alternative 1 . . . J and subscript t indicates the person, then if alternative $j = 1$ is car, $OPTC_{jt}$ takes the value of car-usage costs; if $j = 2$ is bus, then $OPTC_{jt}$ takes the value of bus fares – and so on through the finite set of alternatives J. In each case, for each person the alternatives are ranked in the same order and the value of j has the same meaning.

A ranked alternative can be described by either alternative specific or generic variables, or a combination of both. On the other hand, an unranked alternative can only be described by generic variables. In the case of an unranked alternative, $j = 1$ can have a totally different identity for one person from that for any other one. The example of the set of alternative shopping destinations available to people at different locations, quoted above, serves to emphasise this point. But if one of those destinations is common to every person's set of alternatives, then alternative specific variables can be used to describe that one destination. This is frequently the case in an urban area in which the set of alternative local and regional shopping centres varies for people, but the city centre is relevant to all.

If a generic variable takes the same value for all alternatives $(j \epsilon A_t)$ for all people $(t \epsilon T)$ – i.e. for some k: $X_{itk} = X_{jtk}$, $\forall j \epsilon A_j$, $\forall t \epsilon T$ – then it will have no effect on the model, because of the linear specification. In other words, the coefficient θ_k is not identified. For example, if the variable X_{itk} is income, then we have the same term multiplying the numerator and each member of the sum in the denominator, and thus it will cancel out. This means that all the variables in X_{it} must have alternative specific values. This is not to be confused with alternative specific variables; the reference here is to the value a variable assumes and not to the coefficients of that variable. Hence, the socio-economic characteristics S_t can enter the model only when they are transformed to have alternative specific values. This can be done in two ways:

1 Combining them with the X_i variables

$$X_{itk} = g^k (X_i, S_t)$$

where g^k is a finite function (e.g., price divided by income). The function $g^k(X_i, S_t)$ has an alternative specific value, and can be used to define either a generic or an alternative specific variable.

2 Introducing a series of alternative specific variables, each one of which takes the value of the socio-economic variable for a certain alternative and is otherwise zero. For example, an income variable for alternative 1 which equals income for alternative 1 and zero for all other alternatives. This specification is relevant only for ranked alternatives. Furthermore for a given socio-economic variable, such as income, we can introduce only $N-1$ alternative specific variables, where N is the total number of ranked alternatives. An attempt to estimate a model with N alternative specific variables with the same value (e.g., income) will fail due to perfect 'collinearity' (i.e. the likelihood function will have a ridge for multiples of θ). If we introduce such variables for the first $N-1$ alternatives, then the coefficient estimate of an alternative i specific variable is the difference between the actual coefficient of alternative i variable and that of alternative N variable. This can be observed from rewriting equation (3.26) as follows:

$$P(i:A) = \frac{1}{\sum_{j \epsilon A_t} e^{(X_{jt} - X_{it})'\theta}} \qquad (3.27)$$

If the model was originally written with N alternative specific variables with the same value, the exponents in the sum will include differences of the original coefficients, thus only $N-1$ differences can be identified. It is apparent that this also applies to dummy variables which take the value of one and represent a 'pure alternative effect', e.g., a mode specific constant.

3.5.3 *Independence from irrelevant alternatives property*

'Independence from irrelevant alternatives' (Luce, 1959) implies that the comparison between two alternatives is not influenced by what other alternatives are available. In probability terms the idea is that the ratio of the probability of choosing one alternative to the probability of choosing any other is independent of the set of available alternatives. To show that this is a property of the logit model, consider a subset of A_t consisting of two alternatives i and j. Then, from (3.23) we can write the following ratio:

$$\frac{P(i:A_t)}{P(j:A_t)} = \frac{e^{U_{it}}}{e^{U_{jt}}} = e^{U_{it} - U_{jt}} \qquad (3.28)$$

Hence, the ratio of the probabilities is only a function of the characteristics of the two alternatives. This implies, for example, that if a logit model is used for a mode-choice model, then the ratio of the choice probabilities of, say, car and bus is independent of whether or not rapid transit is also available.

Define Q_{ij} as the odds of choosing i over j:

$$Q_{ij} = \frac{P(i:A_t)}{P(j:A_t)} \qquad (3.29)$$

Then from (3.28) we get:

$$\ln Q_{ij} = U_{it} - U_{jt} \qquad (3.30)$$

If the utility functions are linear, as in (3.26), we get the following linear function:

$$\ln Q_{ij} = (X_{it} - X_{jt})'\theta \qquad (3.31)$$

Thus, the natural logarithm of the odds is expressed as a linear function of the differences of the variables. If the kth variable is alternative i specific, then in equation (3.31) we get:

$$X_{itk} - X_{jtk} = X_{itk}$$

Although this property has frequently been considered desirable, under certain circumstances it could present difficulties in application of the model for forecasting purposes, since, contrary to possible expectations, the introduction of a new alternative would reduce the share of all the other alternatives by equal proportions. Thus, for example, if the share car:bus is currently 60:40 and a new metro is introduced, taking a 25 per cent share of the total market, then it will apparently take 25 per cent from both car and bus, giving a total market split of 45:30:25, car:bus:metro. This problem can be compounded if the new alternative is not a truly independent alternative (Becker et al., 1963; CRA, 1972). In practice most of these problems can be overcome, however, by prior removal or combination of some alternatives, if they are not truly independent alternatives, or by the use of market segmentation.

Thus, as an example of the first, if an express bus service is introduced, then the normal, slower bus service should be disregarded for those trips for which it is dominated by the express service, since the two different types of service cannot be viewed as truly independent alternatives. In the first example quoted above, in reality it could be expected that the metro would attract a higher proportion of bus passengers than car users. If the market were to be segmented into car owners and non-car owners then such a result could be obtained, provided, of course, that the choice probabilities for the two segments were different. The application of market segmentation has been described by Cheslow (1973) in relation to the Northeast Corridor Study.

3.5.4 *Elasticities of the logit model*

A direct elasticity is defined as the percentage change in the dependent variable for a 1 per cent change in an independent variable. In terms of a probability model, it is thus the percentage change in the probability of choosing a given alternative i due to a 1 per cent change in one of the variables in the utility function of that particular alternative. The direct elasticity can therefore be represented thus:

$$E^{P(i:A_t)}_{X_{itk}} = \frac{\partial P(i:A_t) \ / \ P(i:A_t)}{\partial X_{itk} \ / \ X_{itk}}$$

$$= \frac{\partial P(i:A_t)}{\partial X_{itk}} \ \frac{X_{itk}}{P(i:A_t)} \qquad (3.32)$$

Since the probability of choosing one alternative, *i,* is a function of the utility of each of the other alternatives, it is also possible to calculate the percentage change in the probability of choosing a given alternative *i* resulting from a 1 per cent change in one of the variables in the utility function of a second alternative X_{jtk}. This elasticity is described as cross elasticity and can be represented thus:

$$E^{P(i:A_t)}_{X_{jtk}} = \frac{\partial P(i:A_t) \ / \ P(i:A_t)}{\partial X_{jtk} \ / \ X_{jtk}}$$

$$= \frac{\partial P(i:A_t)}{\partial X_{jtk}} \ \frac{X_{jtk}}{P(i:A_t)} \qquad (3.33)$$

For the logit model as in equation (3.26), equations (3.32) and (3.33) result in the following relationships:

$$E^{P(i:A_t)}_{X_{itk}} = \left[1 - P(i:A_t)\right] \ \theta_k \ X_{itk} \qquad (3.34)$$

and

$$E^{P(i:A_t)}_{X_{jtk}} = -P(j:A_t) \ \theta_k \ X_{jtk} \qquad (3.35)$$

In general, we can write:

$$E^{P(i:A_t)}_{X_{jtk}} = \left[\delta_{ij} - P(j:A_t)\right] \ \theta_k \ X_{jtk} \qquad (3.36)$$

where $\delta_{ij} = 1$, for $i = j$
$\qquad 0$, for $i \neq j$

If variable X_{itk} enters the model as $ln X_{itk}$ then the elasticities are as given above, divided by X_{itk}.

The direct elasticity, equation (3.34), varies from zero when the choice probability is one, to $\theta_k X_{itk}$ when the choice probability is zero (if a variable enters the model in a *ln* form, the upper limit of the elasticity will be a constant equal to the coefficient θ_k). If X_{itk} is a price variable we expect a negative direct elasticity and therefore θ_k must be negative.

The cross elasticity, equation (3.35), is dependent only on values related to alternative *j* and not to alternative *i,* for which the cross elasticity is computed. This means that the cross elasticities of all alternatives with respect to an attribute of alternative *j* are the same. This property, equal substitutability, can be considered somewhat of a limitation of the logit model because equal substitutability is not necessarily logical in all cases, and is another aspect of the property of the independence from irrelevant alternatives discussed in section 3.5.3.

The cross elasticity varies from zero, when the choice probability of alternative *j* is zero, to $\theta_k X_{jtk}$, when the same choice probability equals one. Since θ_k for a price variable must be negative, the cross elasticity will be positive.

3.5.5 *Aggregate elasticity*

For a given alternative the aggregate elasticity can be calculated, the aggregate elasticity being with respect to the average choice probability, thus:

$$\bar{P}_i = \frac{\sum_{t=1}^{T} P(i{:}A_t)}{T} \tag{3.37}$$

Assuming that $\dfrac{\delta X_{jtk}}{X_{jtk}} = \dfrac{\delta X_{jk}}{X_{jk}}$ for $t = 1, 2, \ldots, T$, where $X_{jk} = \dfrac{\sum_{t=1}^{T} X_{jtk}}{T}$

i.e. that change in variable X_{jk} is in the same proportion for all individuals, then:

$$E^{\bar{P}_i}_{X_{jk}} = \frac{\delta \bar{P}_i}{\delta X_{jk}} \quad \frac{X_{jk}}{\bar{P}_i}$$

$$= \frac{\sum_{t=1}^{T} P(i{:}A_t) \quad E^{P(i:A_t)}_{X_{jtk}}}{\sum_{t=1}^{T} P(i{:}A_t)} \tag{3.38}$$

This (3.38) holds also for equal absolute changes, $\delta X_{jtk} = \delta X_{jk}$, if the value of the variable X_{jtk} is the same for all individuals t. When the change in X_{jtk}, δX_{jtk}, is the same for all individuals t, but the value of X_{jtk} is not the same, then:

$$E^{\overline{P}}_{X_{jk}} = \frac{\sum_{t=1}^{T} P(i:A_t) \; E^{P(i:A_t)}_{X_{jtk}} \; \frac{X_{jk}}{X_{jtk}}}{\sum_{t=1}^{T} P(i:A_t)} \qquad (3.39)$$

The aggregate elasticity is in fact a weighted sum of the individual elasticities using the individual probabilities as the weights. As a general rule, it can be shown that the elasticity of a sum of functions is the weighted sum of the elasticities of the functions with the functions as the weights. It can also be shown that the elasticity of a product of functions is the (unweighted) sum of the elasticities of the functions. These two rules can be used to compute the elasticity of the joint probability in a recursive structure (see, for example, Manheim, 1970).

3.5.6 *Estimation technique*

For a cross-section of behavioural units making a choice, we do not observe the probabilities but only the actual choices. Hence, with disaggregate data the observed dependent variable takes a value of either zero or one. The independent variables are continuous and/or discrete. With observed 0, 1 dependent variables, we use the maximum-likelihood method to estimate the model coefficients. The estimator of the θs of the logit model as in equation (3.26) and its properties are described by McFadden (1968).

If we aggregate groups of behavioural units according to socio-economic categories or geographic location, and use the groups as the observation units, then the observed dependent variable is a share with a value between zero and one. With grouped data both maximum-likelihood and ordinary least-squares techniques can be used. The ordinary least-squares technique is based on the linearisation of the model by dividing the probability of each alternative by a 'base' alternative and taking logs, as in equation (3.31). The estimated coefficients are therefore sensitive to the choice of a 'base' alternative, except for binary choice (see, for example, McLynn and Woronka, 1969). A maximum-likelihood estimator with aggregate data is, in principle, a simple extension of the disaggregate estimator.

The likelihood function of a disaggregate sample is written as follows:

$$L = \prod_{t=1}^{T} \prod_{i \epsilon A_t} P(i:A_t)^{g_{it}} \tag{3.40}$$

where T is the number of observations and g_{it} equals one if alternative i was chosen in observation t, and zero otherwise. Taking the logarithm of both sides we get:

$$ln L = L^* = \sum_{t=1}^{T} \sum_{i \epsilon A_t} g_{it} \quad ln P(i:A_t) \tag{3.41}$$

Substituting equation (3.26) for $P(i:A_t)$, we get for the first order conditions:

$$\frac{\delta L^*}{\delta \theta_k} = \sum_{t=1}^{T} \sum_{i \epsilon A_t} \left[g_{it} - P(i:A_t) \right] X_{itk} = 0 \tag{3.42}$$

for $k = 1, \ldots, K$.

This is explained in greater detail by McFadden (1968). The K equations in (3.42) are non-linear and their solution requires an iterative procedure. In this study we used the Newton–Raphson method. The estimation computer program used was developed by C.F. Manski at the Department of Economics at MIT.

McFadden (1968) showed that, except under very specific conditions in the data, the maximum of L^* obtained from (3.42) is unique. This estimator has optimal asymptotic properties; as the number of observations increases, it can be shown that the estimator approaches 'optimality', i.e. approaches minimum variance in the estimate of each coefficient. The asymptotic variance-covariance matrix is the inverse of the matrix of second derivatives of L^*, multiplied by minus one (Theil, 1971).

It should be noted that, since we do not observe probabilities, it will be misleading to compare the computed probabilities with the g_{it} variables, if we assume that the actual choice is made with a probability and not with certainty, as the g_{it} variables would indicate. Therefore, a 'goodness of fit' measure, such as R^2 in ordinary least-squares, which is based on estimated residuals, is not appropriate. In addition, a comparison of a sum of probabilities for a given ranked alternative with the total number of observations in which this alternative was chosen, is also misleading under the following conditions. If the set of variables includes an alternative specific constant:

$$X_{tjk} = \begin{cases} \text{constant, for } j = i \\ 0, \quad \text{for } j \neq i \end{cases} \tag{3.43}$$

then from the first order conditions, equation (3.42), the following always holds:

$$\sum_{t=1}^{T} g_{it} = \sum_{t=1}^{T} P(i:A_t) \tag{3.44}$$

Thus, for a disaggregate logit model, because we cannot estimate residuals, there is no R^2 statistic which will indicate how well the model 'fits' the data. However, it is possible to define a measure analogous to R^2, which is based on the value of the *ln* likelihood function, and can be used to compare alternative models, as follows:

$$\rho^2 = 1 - \frac{L^*(\hat{\theta})}{L^*(0)} \tag{3.45}$$

where $L^*(\hat{\theta})$ is the value of L^* for the vector of estimated coefficients and $L^*(0)$ is the value of L^* for $\theta = 0$. Since the likelihood function is the product of probabilities, and the probabilities are always between zero and one, the likelihood function must be between zero and one, and the *ln* of the likelihood function must always be negative. Therefore, the process of maximising likelihood is to raise $L^*(\hat{\theta})$ from a large negative number, $L^*(0)$, to as close to zero as possible. Thus, ρ^2 is equal to the ratio of the explained *ln* likelihood over the total *ln* likelihood, and it lies between zero and one.

This measure suffers from the same deficiencies as R^2. In particular, it does not take account of the degrees of freedom. Adjusting for these, we get:

$$\bar{\rho}^2 = 1 - \frac{L^*(\hat{\theta}) \Big/ \sum_{t=1}^{T} (J_t - 1) - K}{L^*(0) \Big/ \sum_{t=1}^{T} (J_t - 1)} \tag{3.46}$$

where J_t is the number of alternatives in A_t and K is the total number of variables specified.

Setting $\theta = 0$ amounts to an assumption of equally likely alternatives. When all utility functions go to zero the probability of any alternative goes to $1/J_t$ since $e^0 = 1$. However, for ranked alternatives we can compute a version of $L^*(0)$ that takes into account pure alternative effects, i.e., sets initial estimates of the probabilities equal to the observed frequencies using the alternative specific constants.

3.5.7 *Other properties of this estimator*

Consider the simultaneous model in which we estimate directly $P(m,d{:}DM)$, and a recursive structure in which we estimate separately two probabilities, say $P(d{:}D)$ and $P(m{:}M_d)$. The problem is how to compare the goodness of fit between the recursive and the simultaneous structures. For the simultaneous model we can compute ρ^2 directly. For the recursive model we have to compute the joint likelihood of $P(d{:}D)\ P(m{:}M_d)$ in order to be able to calculate a comparable ρ^2. The joint *ln* likelihood of the recursive model can be written as follows:

$$L^*_{dm} = \sum_{t=1}^{T} \sum_{dm\epsilon DM} g_{dmt} \quad ln \left[P(d{:}D_t) \quad P(m{:}M_{dt}) \right]$$

$$= \sum_{t=1}^{T} \sum_{d\epsilon D_t} \sum_{m\epsilon M_{dt}} g_{dmt} \quad \left[ln\ P(d{:}D_t) + ln\ P(m{:}M_{dt}) \right]$$

$$= \sum_{t=1}^{T} \sum_{d\epsilon D_t} g_{dt} \quad ln\ P(d{:}D_t) + \sum_{t=1}^{T} \sum_{m\epsilon M_{dt}} g_{d'mt} \quad ln\ P(m{:}M_{d't})$$

$$= L^*_d + L^*_{m|d} \tag{3.47}$$

where d' is the chosen destination in observation t, and L^*_d and $L^*_{m|d}$ are the *ln* likelihood values from the two separate models. Thus, the joint *ln* likelihood function of a recursive model is equal to the sum of the *ln* likelihood functions from the structural probabilities.

An asymptotic t test can be used to test the significance of a single coefficient. The significance of a group of coefficients can be tested using the statistic of minus twice the *ln* of a likelihood ratio. This statistic is asymptotically distributed as chi-square (Theil, 1971).

3.6 Application of disaggregate models: the aggregation problem

Whilst we consider that travel demand models can be estimated more satisfactorily (i.e. better models can be estimated using the same basic set of data) and more economically using disaggregate data, disaggregate models are not directly applicable for normal travel demand predictions, since many of the necessary socio-economic data are not available in disaggregate form.

The basic problem with the use of disaggregate models in the prediction of aggregate travel behaviour is the development of a procedure for expanding individual choice estimates over the population of interest to obtain a reliable, unbiased forecast of group behaviour. The transformation of a disaggregate model to an aggregate one is a complex problem, and has been reviewed by Koppelman (1974) and Westin (1974).

If the disaggregate model is linear, then the aggregate model has the same linear specification with averages of the variables substituted for the individual values. However, if the disaggregate model is non-linear, the disaggregate functional specification, with averages of the independent variables substituted for individual values, will give a biased forecast of the average of the dependent variable. This is demonstrated by considering a simple binary mode-choice model describing the probability of a selected trip-maker choosing public transport. The assumption that parameters are identical for each individual (in the market of interest) will be used throughout.

$$P_t = f(U_t)$$

and
$$U_t = \theta_0 + \theta_1 X_{1t} + \theta_2 X_{2t} + \theta_3 X_{3t} \qquad (3.48)$$

where t denotes an individual $t = 1, \ldots, t, \ldots, T$;
X_{1t} is the difference in travel cost by car and public transport divided by income for individual t;
X_{2t} is the difference in in-vehicle travel time by car and public transport for individual t;
X_{3t} is the difference in out-of-vehicle travel time by car and public transport for individual t; and
θ_0 θ_1 θ_2 θ_3 are coefficients.

If
$$f(U_t) = U_t$$

it can easily be shown that the expected fraction of the market using public transport is

$$\bar{P} = \theta_0 + \theta_1 \bar{X}_1 + \theta_2 \bar{X}_2 + \theta_3 \bar{X}_3 \qquad (3.49)$$

where

$$\bar{P} = \frac{\sum_{t=1}^{T} P_t}{T}$$

and

$$\bar{X}_k = \frac{\sum_{t=1}^{T} X_{kt}}{T} \qquad k = 1, 2, 3$$

In this case, a change in transportation policies which results in a change in travel time by car can be represented by the average value of the change, and can immediately enter the model to predict the effect on the aggregate modal split. The aggregate function, used in this manner, would provide unbiased predictions.

If, however, the function is non-linear, it can be shown that, in general,

$$\bar{P} \neq f(\theta_0 + \theta_1 \bar{X}_1 + \theta_2 \bar{X}_2 + \theta_3 \bar{X}_3)$$

It will be equal only if all trip-makers concerned have identical values for the variables included or, more generally, the same value for U_t.

To illustrate this, assume the following logit formulation:

$$P_t = f(U_t) = \frac{1}{1 + e^{-U_t}} \qquad (3.50)$$

This relationship can be depicted in Fig. 3.1. The influence of a small change in U_t (or any of the X_{kt}) on the probability of using public transport, i.e. the slope of the curve in Fig. 3.1, depends on the prior value of U_t. If we consider a population with $\bar{U}$ represented by the point B and assume that all $U_t = \bar{U}$, the change in the aggregate modal split will be the change in P_t at B. However, if the true population is made up of two subgroups represented by points A and C, where responsiveness to change in U_t is much lower than at B, the change in the aggregate modal split predicted by use of $\bar{U}$ as a representative value will be biased upward.

Consider a group of people with differing socio-economic attributes.

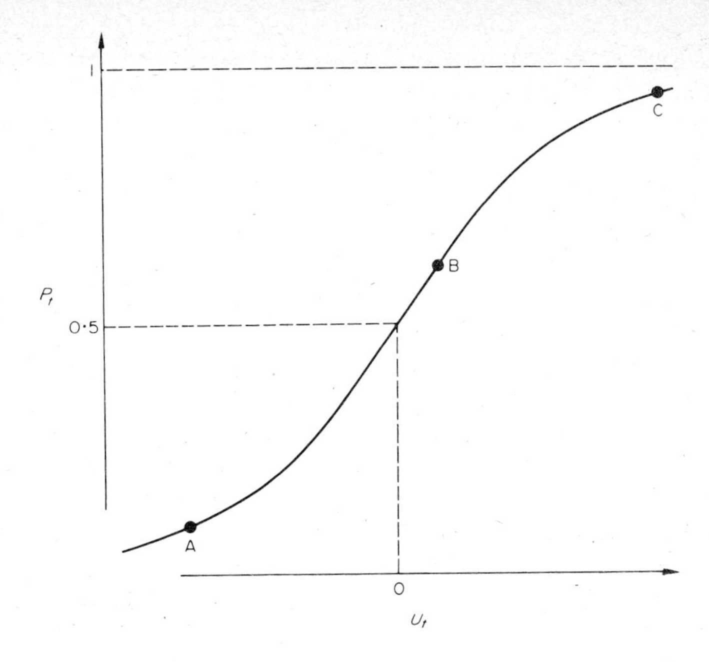

Fig. 3.1 Diagrammatic representation of the logit model

Each person within this group has to choose one alternative from a set of mutually exclusive and collectively exhaustive alternatives, each alternative having a set of attributes. In a mode-choice model these alternatives are the various travel modes. Further, assume that the distribution of socio-economic variables, S, and attributes of all alternatives, X, is known. Denote this joint distribution as $f(X,S)$. If the probability of a person with characteristics S choosing an alternative i when confronted with alternatives described by attributes X is denoted as $P(i{:}X,S)$, then the following equation holds:

$$\overline{P}_i = \int\int_{XS} P(i{:}X,S)\, f(X,S)\, dS\, dX \tag{3.51}$$

where $\overline{P}_i$ is the expected fraction of total population choosing alternative i.

Thus the transformation of a disaggregate model into an aggregate model could, in principle, be performed by integrating the relationship over the distribution of the independent variables. This means that, for a non-linear model, the aggregate model will be specific to the population and area, the distribution of whose characteristics was used in the integration procedure. This means that an aggregate model estimated with aggregate data will be valid for prediction only if the distribution of the population in each zone under study is similar to that in the zones for which the model was estimated. Furthermore, changes in the distribution of these independent variables will not be reflected in the forecast produced by the aggregate model.

On the other hand, disaggregate models do not include any implied distribution of variables. Therefore they can be aggregated and applied over any population or area of interest. In addition, their predictions will be sensitive to changes in the distribution of the independent variables.

The transformation by integration of a disaggregate non-linear model into an aggregate model can be an intractable mathematical problem. At best, it requires considerable effort to develop all the necessary independent variable distributions. However, facing this problem head-on would appear to be more desirable than assuming it away, which the more expedient aggregate models in effect do.

To address further the problem of dealing with the independent variable distributions in forecasting with a disaggregate model, in any realistic forecasting situation it is extremely unlikely that the joint distribution of attributes and socio-economic variables will be known. Thus, a number of simplifications are generally introduced to make the task of forecasting feasible.

First, it is possible to assume that all, or some, of the attributes and characteristics are distributed independently. Thus:

$$f(X,S) = f_1(X_1,S_1)\ f_2(X_2,S_2)\ f_3(X_3,S_3)\ldots \tag{3.52}$$

In the simplest case, the distribution of each attribute and characteristic is assumed to be independent of all others. Hence:

$$f(X,S) = f_1(X_1)\ f_2(X_2)\ldots f_L(X_L)\ f_{L+1}(S_1)\ldots f_{L+K}(S_K) \tag{3.53}$$

Using a mode-choice model as an example, it might be assumed that the income distribution of a market segment is independent of the distribution of the level of service.

A further simplifying assumption which can be used is to require that the distributions follow a known form. For example, income is often modelled using a gamma distribution. This assumption reduces the problem of estimating and forecasting an entire distribution to one of predicting a fairly small set of parameters.

A naive simplification of the aggregation problem is to use only an estimate of the means of the distribution of attributes and socio-economic characteristics, which is what is done with aggregate models. This reduces the general aggregation problem to one of simply predicting at the mean. Thus, if $\overline{X}$ and $\overline{S}$ are the vectors of mean attribute value and mean socio-economic characteristics respectively, the fraction choosing $i = P(i{:}\overline{X},\overline{S})$.

It is possible to use various combinations of these assumptions in forecasting. For example, in predicting modal split for a zone, one could use the mean level of service values in that zone but might use a gamma function to describe the income distribution, assuming income to be independent of other variables.

An approximation method which is useful because of its simplicity is based upon use of relatively homogeneous groups of people, or market segments. First, the population is classified into market segments according to the most important variables in the model, and then for each market segment we substitute the average of X and S in the model separately.

Any aggregation method requires the forecast of some distributions of critical characteristics in addition to their average values. However, even in the case for which no forecast distributions are available, the use of existing distributions for zones of interest will lead to aggregate results that must be better than those based on aggregate models, since the latter imply some average distribution which is not zone specific. Furthermore, when the aggregation is explicit, judgement may be used in a Bayesian sense to suggest modifications in the distribution used.

4 Study Design and Data

4.1 Urban transportation in the Netherlands

The urban transportation scene in the Netherlands is characterised by the number of modes available and in common use: bus, car, bicycle, moped and foot all play a significant part. In the larger conurbations train is also utilised for intra-urban trips and in Amsterdam, The Hague, Rotterdam and Delft there is also an extensive tram network, supplemented in Rotterdam by a metro.

In 1966 9·4 per cent of all week-day trips in the west of the country were made by public transport (Table 4.1), with bus and tram accounting for 7·6 per cent and train 1·8 per cent (COVW, 1970). While 10·7 per cent of all trips made within the four major conurbations were by bus or tram, only 0·8 per cent of all week-day trips made in the other urban areas were by public transport. From Table 4.1 it is quite evident that in these latter areas the bicycle and moped were predominant, accounting for 55·6 per cent of all week-day trips; even within the four conurbations, bicycle and moped were used for 35·7 per cent of all trips.

Whilst the bicycle is a well-established and well-known Dutch phenomenon, the moped is more recent, the number in use having doubled between 1960 and 1971. The moped is a motorised bicycle with an engine of not more than 50cc capacity; it can be ridden by anyone aged 16 years or over who is in possession of third-party insurance. Officially mopeds are limited to 30km per hour within built-up areas and 40km per hour elsewhere; in practice these speed limits are not enforced, with the result that, combined with their high degree of manoeuvrability, journey times considerably shorter than those of either bus or car can be achieved under congested urban conditions. Passengers are allowed to be carried, and the Dutch Central Bureau of Statistics estimates that in 1971 mopeds were used for 9,770 million passenger kilometres, compared with 78,400 million for private cars (CBS, 1973). The number of mopeds in use in 1971 was 2,100,000.

Faced with increasing competition from both the private car and the moped, with population growth taking place on the edges of existing built-up areas as well as within villages, public transport has suffered a relative decline in patronage (Fig. 4.1). Nevertheless, the increase in the

Table 4.1

Modal split in the west of the Netherlands, week-day 1966 (COVW, 1970)

Mode	Percentage share per mode						
	Within the four conurbations*	Within other urban areas	Between the four conurbations *, and between the four conurbations * and other urban areas	Between other urban areas	Between all urban and rural areas	Between and within rural areas	Total
Foot, 10 minutes or more	28·6	23·4	1·3	2·6	1·0	16·7	22·1
Bicycle	25.4	44·5	6·5	24·8	17·6	45·1	30·3
Moped	10·3	11·1	7·3	14·1	11·2	11·7	10·7
Private car	21·3	16·6	44·9	36·7	41·1	18·0	23·1
Tram, bus	10·7	1·8	10·6	8·6	10·3	1·2	7·6
Train	0·2	0·0	20·9	4·5	5·8	0·2	1·8
Taxi and coach	1·5	1·0	4·2	4·9	5·6	1·9	1·9
Other	2·0	1·6	4·3	3·8	7·4	5·2	2·5
Total	100·0	100·0	100·0	100·0	100·0	100·0	100·0

* Amsterdam, Rotterdam, The Hague, Utrecht.

Dutch population, as well as increasing mobility, have resulted in little change in the number of passenger kilometres carried by public transport in the last decade (Fig. 4.2). Most larger urban areas have therefore been able to maintain their bus networks, although in recent years only through receipt of government subsidies. However, as elsewhere in the Western world, the problems associated with the motor car, urban highway construction costs, environmental impact, noise and air pollution, and

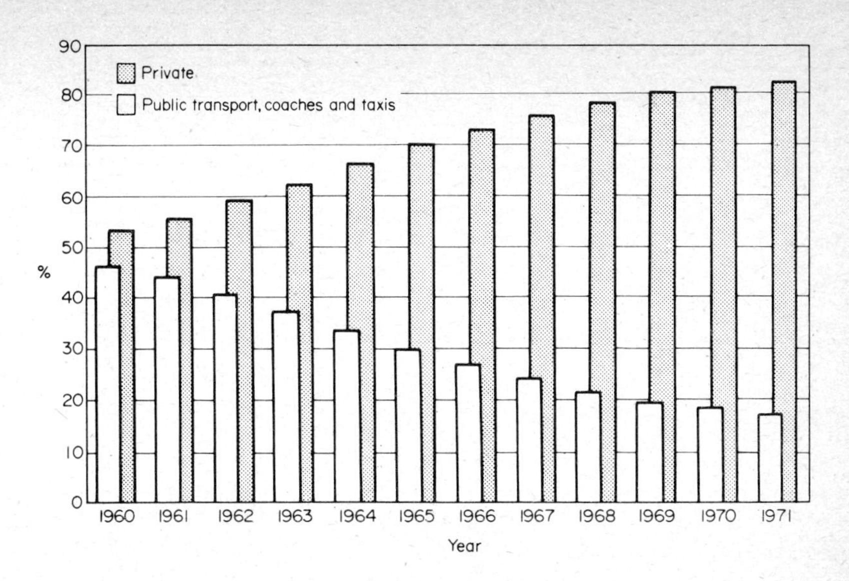

Fig. 4.1 Percentage division of passenger kilometres over private and public transport, the Netherlands, 1960–71

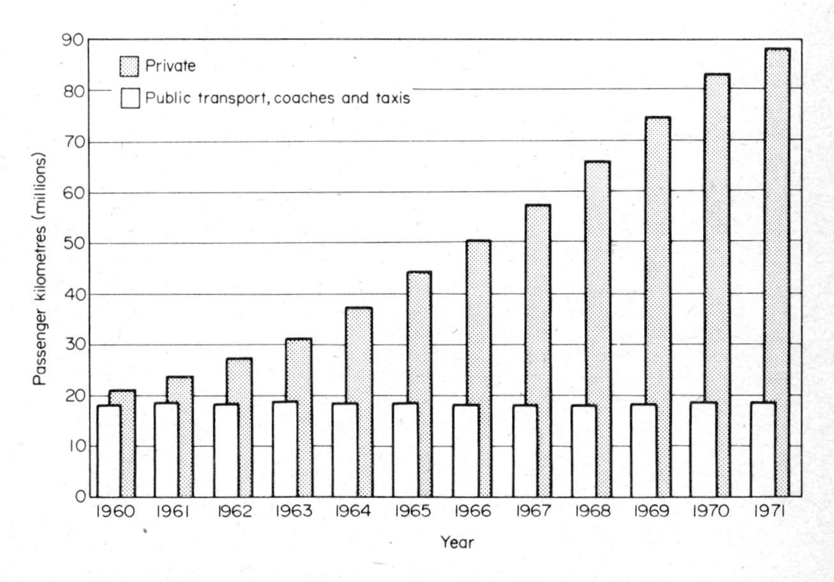

Fig. 4.2 Passenger kilometres by private and public transport, the Netherlands, 1960–71.

most recently the energy crisis, have turned people's attention back towards public transport and its potential role in urban life.

Car ownership, however, continues to grow, having doubled between 1966 and 1971, when it was estimated that there were 208 cars per 1,000 people. Whilst forecasts of future levels of car ownership vary considerably, the national transportation study completed in 1972 forecast 410 cars per 1,000 persons by 1996 (NEI, 1972); an alternative forecast is of a saturation level of 300 cars per 1,000 persons (Hupkes, 1974).

Past trends and policies have, of course, been based upon low cost, freely available sources of energy, a scenario which has suddenly changed. Those concerned with the planning of our communities, and the utilisation of existing investments, are faced with a new set of problems, new issues. In the resolution of these problems and the development of strategies for the future, techniques that provide a better understanding of the behaviour of people, and of the sort of considerations that influence a decision are necessary. Through such an understanding it should be possible to significantly improve the manner in which both the development of new communities and changes within existing communities are guided.

4.2 VODAR

In the spring and early summer of 1970 a home-interview survey, code-named VODAR, was conducted in Eindhoven and four adjacent municipalities (Buro Goudappel en Coffeng, 1971; Richards *et al.,* 1972). This survey was sponsored by the Rijkswaterstaat (the central government agency in the Netherlands with responsibility for urban transportation studies) and carried out by Buro Goudappel en Coffeng. The purpose of the survey was two-fold. One was to compare different procedures then in use in urban transportation studies in the Netherlands; the second was the establishment of a data-base for the development and evaluation of techniques for the analysis and prediction of urban travel demand. Data were obtained in the course of the survey from 14,000 households (one in five), distributed throughout the study area. These data related to the socio-economic characteristics of both the individual survey respondents and their households, as well as trips made by each respondent (aged five years and over) during a specified 24-hour week-day period.

The data collected in the survey were to be used for a number of different purposes, many of which had, or could be expected to have, different requirements for the geo-coding of the trip origins and destinations. A geo-coding system had therefore to be devised which would provide the maximum of flexibility in the potential future applications of the data. After an extensive study of possible systems, it was decided to use a rectangular co-ordinate referencing system. For trip ends within the study area a 10 metre grid was utilised; for trip ends outside the study area a 1 km grid was applied, with some local exceptions where a 500 metre grid was applied. The geo-coding system within the study area was thus extremely fine, giving an accurate location of origins and destinations. It is thought that this is probably one of the finest geo-coding systems ever applied to such a large quantity of urban transportation survey data. Such a system is clearly ideally suited for application in geographically disaggregate models. In fact, this is one of the few disaggregate model studies undertaken with disaggregate geographic data, the majority of studies to date having had to accept a higher level of geographic abstraction.

The socio-economic and trip data collected in the course of the survey were:

1 *Household data:* age and sex structure,
vehicle ownership (bicycles, mopeds, motor-cycles and scooters, cars, and vans),
gross household income.

2 *Person data:* age,
sex,
position in household,
type of driving licence held,
economic activity,
number of hours per week in paid employment,
type of employment,
occupation,
industry in which employed.

3 *Trip data:* address of trip origin,
time of departure,
land use at origin,
purpose at origin,
address of trip destination,
time of arrival,
land use at destination,
purpose at destination,
mode.

The survey data were subjected to an extremely thorough validation and correction procedure, both manual and by computer (Buro Goudappel en Coffeng, 1974); they were thus in a state suitable for immediate application in further studies.

4.3 The VODAR study area

Eindhoven is a modern industrial town in the south of the Netherlands (Fig. 4.3). Although only a small town at the turn of the century, it was in Eindhoven that G. Philips established his first factory, and the town has

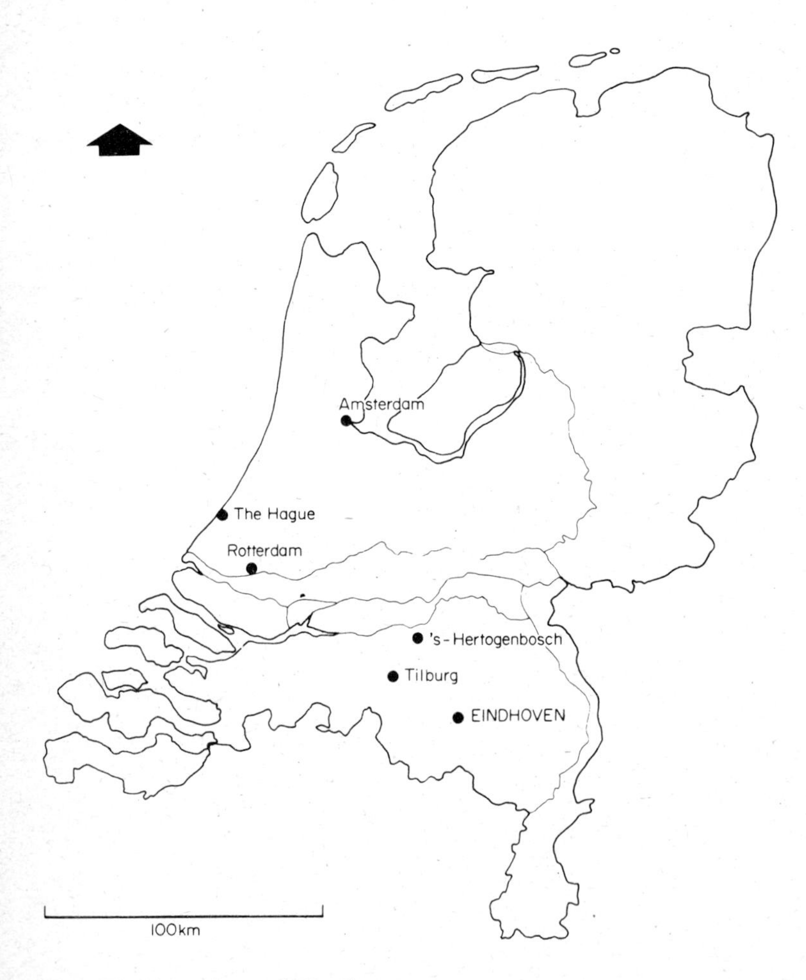

Fig. 4.3 The location of Eindhoven

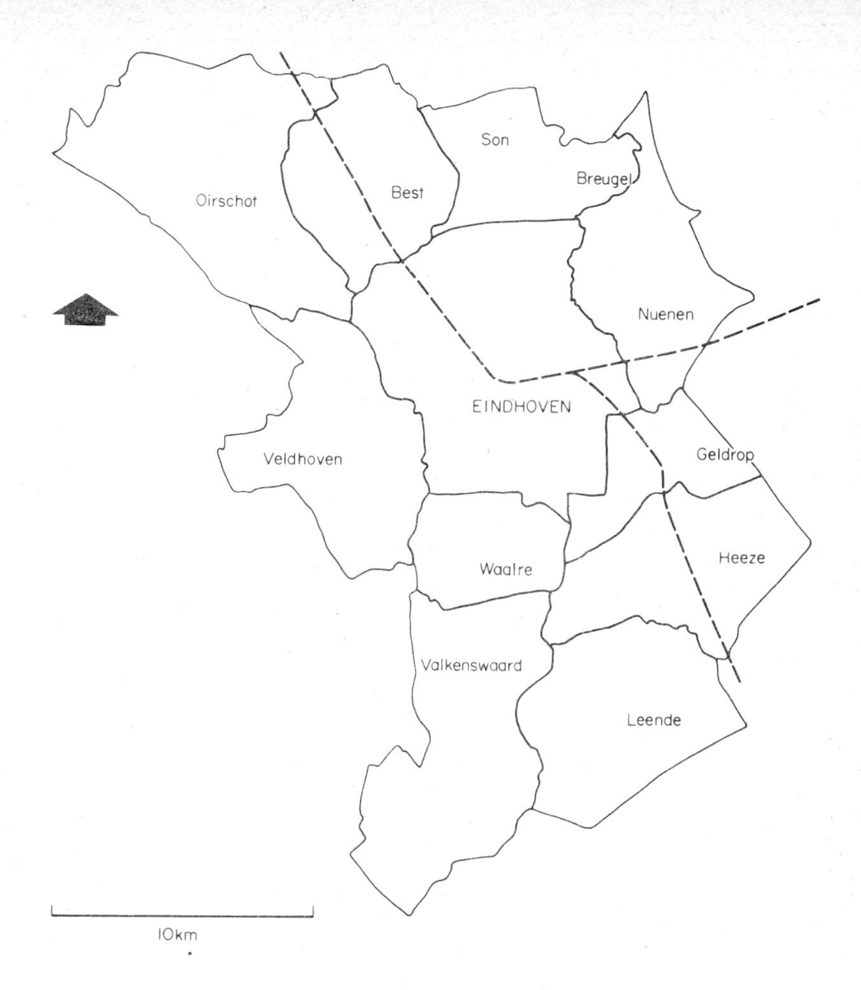

Fig. 4.4 Agglomeratie Eindhoven

grown with the rise of Philips as one of the world's electrical giants. Today Eindhoven is the site of Philips' international headquarters as well as their main Dutch research and manufacturing centre. Eindhoven has a population of 190,000 and is the centre of a group of eleven municipalities, Agglomeratie Eindhoven (Fig. 4.4), with a total population in 1972 of 345,000. In addition to Philips, Eindhoven is also the home of the Netherlands' only native car and truck manufacturer, DAF. DAF employ some 7,000 people in their Eindhoven plant; Philips employ some 36,000 people in the Agglomeratie. The total number of jobs within the Agglomeratie was estimated to be 135,000 in 1970. Although not a one-industry town, Eindhoven is clearly dependent to a very great extent, directly and indirectly, on its two major employers.

The VODAR study area, the built-up areas of the municipalities of Best, Eindhoven, Geldrop, Son and Breugel, and Veldhoven, embraced some 80 per cent of the 1970 population of the Agglomeratie. The five municipalities within the VODAR study area are all essentially different from each other.

Best (1970 population 16,500) is an industrial township of relatively low-income households (Fig. 4.5). The centre of Best lies some 10km to the north-west of the centre of Eindhoven and the two built-up areas are separated by 3km of primarily open country. Although there are a number of small, neighbourhood type shopping centres within Best, there is no large compact shopping centre. There is, however, one group of shops, including a small supermarket and three banks, close to the town hall, which could be described as the 'town centre'. Best lies on the main railway lines between Eindhoven and the north ('s-Hertogenbosch, Utrecht, Amsterdam) and the west (Tilburg, Rotterdam, The Hague) with a half-hourly service on the line to the north and an hourly service on that to the west. There is also an hourly bus service between Tilburg and

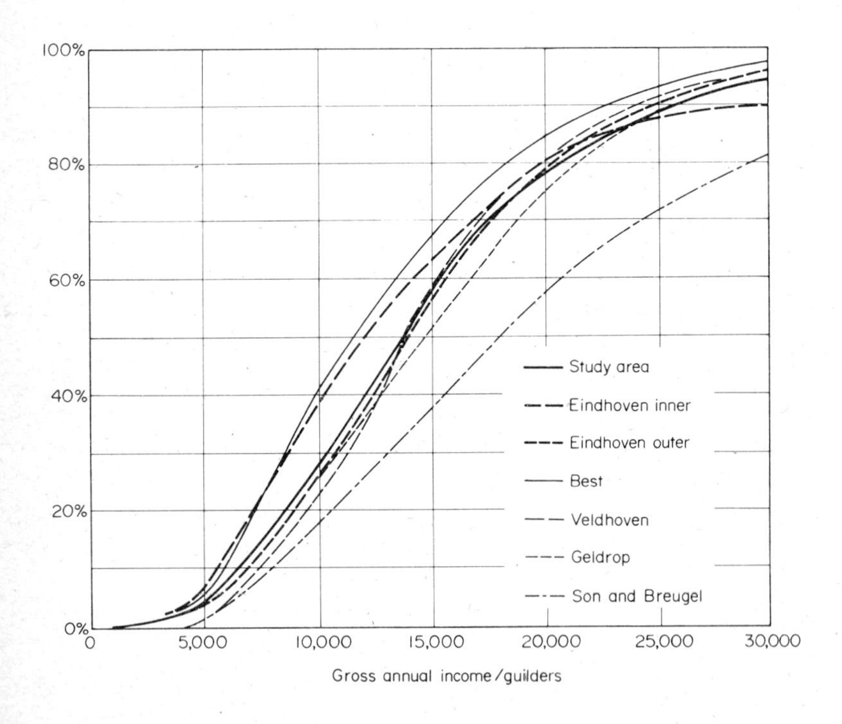

Fig. 4.5 Household income distribution, VODAR study area

Eindhoven which passes through Best, and a half-hourly service between 's-Hertogenbosch and Eindhoven. There is no internal bus service within Best, but the long distance buses are so routed that they can be used for some internal trips. In fact very little use is made of them for such trips.

Son and Breugel is a small municipality (1970 population 10,800) comprising two main villages, Son and Breugel, which over the years have grown together to form one urban area. It is primarily a medium- and upper-income residential area with only limited local employment. Shopping facilities are spread throughout the area, although there is a minor concentration in the vicinity of a medium sized branch of Albert Heijn, a national chain of supermarkets. Son and Breugel is situated 10km to the north of Eindhoven, from which it is separated by a strip of semi-developed countryside some 2km wide. Son and Breugel are linked to Eindhoven by a half-hourly bus service; there is no internal bus service. Neither Son nor Breugel is on a railway line.

Veldhoven, the second largest municipality within the Agglomeratie (1970 population 27,300), lies immediately to the west of Eindhoven and the two built-up areas are almost contiguous. Whilst Veldhoven functions to a very great extent as a dormitory town for Eindhoven, there is some local industrial employment. The distribution of household incomes is very similar to that of the VODAR study area as a whole, but with rather fewer low-income and more high-income households (Fig. 4.5). There is no internal bus service nor is there any railway line in the vicinity. Two half-hourly bus services provide a 15-minute interval service between Veldhoven and Eindhoven. The one service links Eindhoven and Belgium; the other, which terminates in Veldhoven, follows a more circuitous route.

Geldrop (1970 population 26,900) lies immediately to the east of Eindhoven, to which it is joined by a strip of ribbon development. The DAF plant, although mainly in Eindhoven, has been extended across the municipal boundary into Geldrop. Like the other three municipalities, Geldrop is highly dependent upon Eindhoven, although there is some industrial employment. Geldrop is essentially a medium-income area (Fig. 4.5). The main railway line from the west of the Netherlands through Eindhoven to the south-east of the country, South Limburg, passes through Geldrop, and there is a half-hourly rail service between Eindhoven and Geldrop. Geldrop is linked by four different bus routes to Eindhoven, providing a 15-minute interval service between the two centres.

Eindhoven, the fifth largest town in the Netherlands, is a major regional service centre, serving an area much larger than the Agglomeratie. Its shopping centre is modern and includes one of the four branches of the Netherlands' quality department store, the Bijenkorf. In addition to being the home of Philips and DAF, Eindhoven boasts a large modern technical university, one of three in the Netherlands. As might be expected of a modern town, the highway network is well developed, and, outside the town centre, congestion is limited. There is also a local bus service which provides relatively good coverage of most of the residential areas (Fig. 4.6). The normal frequency on most routes is 30 minutes, although on some routes there is a 15-minute frequency. The combination of routes as they approach the centre effectively provides many parts of the town with a 15-minute service. Although patronage remained fairly constant

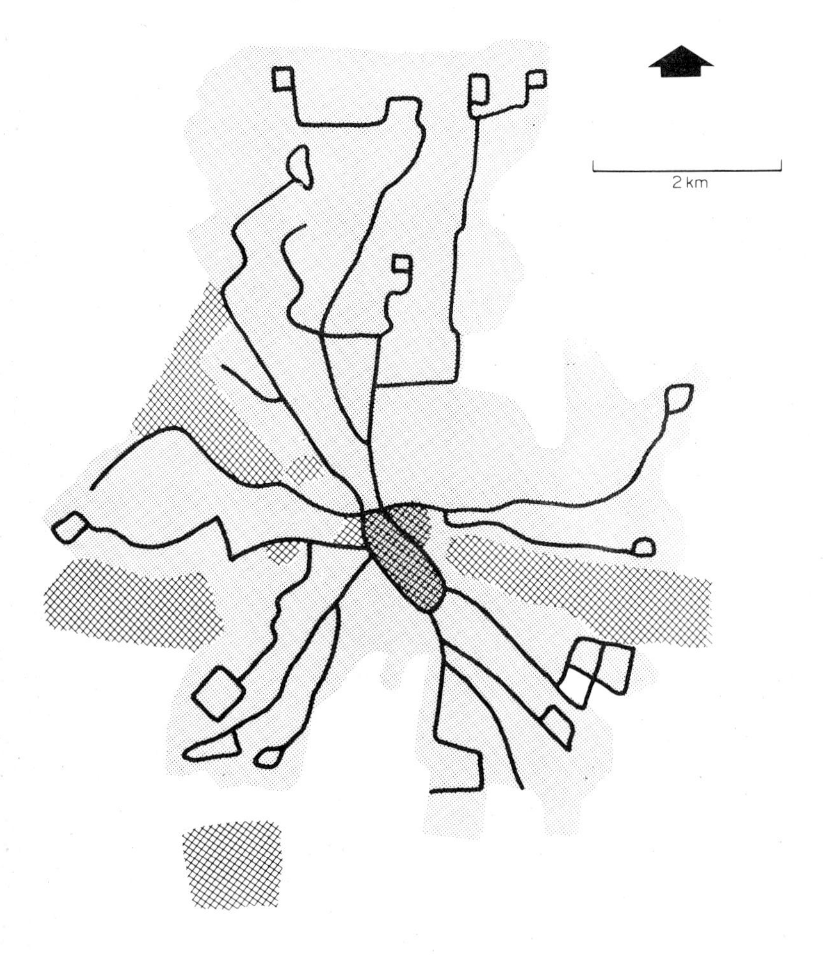

Fig. 4.6 Local bus network, Eindhoven

from 1966 to 1970, at between 11·5 and 12·0 million passengers per year, it has since fallen sharply – only 8·5 million passengers were carried in 1972.

4.4 The project strategy

The basic study objective, an evaluation of disaggregate and simultaneous techniques applied to passenger travel demand modelling, is clearly very broad. In order to achieve any useful results within the available study budget, some more limited goals were necessary. The data available from the VODAR study provided in themselves only limited constraints, within the sort of modelling strategy common to traditional urban transportation studies. The greatest constraint, in so far as data were concerned, was the availability of level of service data. In order to realise the full potential of the geographically disaggregate origin and destination data available in VODAR, level of service data had to be determined for each relevant alternative for each observation. For work trips this involved the estimation of up to five sets of modal data per observation, for shopping up to 55. Furthermore, no level of service data, including parking location, were available from the VODAR files.

In order to minimise the complexity of the project it was decided to focus the work initially on a part of the total trip decision spectrum in which a simple choice could be considered to exist. That part which most nearly satisfies this demand is of mode-choice for trips between home and a regular place of work. Given the relatively low job mobility of the Dutch, and the manner in which public housing is allocated (only a minority of the population live in houses that they own themselves), it can be expected that the choice of mode for the journey to work is, in the short run, independent of the choice of either work place or residential location for the vast majority of the working population.

For the evaluation of a simultaneous model, a relatively simple situation was also sought. The modelling of frequency, time of day, destination-choice, mode-choice and route-choice simultaneously, with the current level of experience, was likely to pose so many fundamental problems that no useful results could be expected within the context of this study. Simultaneous modelling of destination-choice and mode-choice appeared to offer better prospects than any other combination; furthermore, the availability of trip data for only a 24-hour period made estimation of trip frequency at the disaggregate, or household, level difficult. Indeed, for a purpose such as shopping it did not seem totally unreal to assume that the decision to make a shopping trip was less dependent upon the alternative destinations and modes available than was the choice of destination upon the alternative modes available, and *vice versa.* Analysis of all the non-work trip purposes in which a real interdependence between destination- and mode-choice could be considered to exist (thus trips to and from places of education were excluded), as well as analysis of the problems associated with the identification and quantification of alternative destinations, led to the choice of shopping trips for this element of the study.

4.4.1 *The home–work–home mode-choice model*

In order to keep the situation being modelled as simple as possible, it was decided to model simple home–work–home chains in which the mode for both legs forming the complete chain (or journey) was the same. The choice of mode for one leg of a home–work–home chain can be expected to be influenced by different considerations if the mode chosen is different to that chosen for the second leg than if the same mode is utilised for both legs. This is also true of mode-choice for trips between work and home (or *vice versa*) which form part of a three-or-more-leg journey (a journey being defined as a series of trips beginning and terminating at home). Furthermore, people who go home at lunch time and then return to work could be expected to have a different set of values from those remaining at work all day; a division of the survey population into those going to work in the morning and returning home in the evening (single chains), and those also returning home at lunch time (double chains) therefore also seemed to offer a reasonable basis for model design.

Although eleven modes were utilised in the collection of data, there were few observations for four of these (truck and delivery van, taxi, motor-cycle and scooter, and non-scheduled bus and coach). Furthermore, the choice mechanism leading to the selection of some of these modes (if indeed there was any real choice), and including that of car passenger, could be expected to be different from that applied when one of the remaining modes was selected. Thus, in view of the objective of relative simplicity, it was decided that only the following modes would be considered in the model: car (driver), moped, bicycle, train, bus (scheduled service), and foot.

Whilst the model specification was thus simplified, the volume of data available from the VODAR study was too great for all valid observations to be utilised; for the purposes of this project a sample had to be selected.

This sample could have been selected at random from the total set of data, but this would have involved more work in the preparation of the necessary level of service data than for a clustered, geographic sample. Furthermore, a sample for selected geographic areas would facilitate the evaluation of the geographic stability, or transferability, of the estimated model. Examination of the study area, in combination with a data requirement of between 1,000 and 1,500 households, led to the selection of Best, and Son and Breugel. Best has the lowest income of the whole study area, has a relatively high proportion of local employment and has a train service connecting it with Eindhoven. Son and Breugel is a medium- and high-income area, has little local employment and no train service. Both are a similar distance from Eindhoven. These two municipalities, from which data for some 1,300 households were available, thus covered a large part of the complete income spectrum. The sample could be expected to contain short and long journeys to work, as well as respondents for whom train was an alternative mode and respondents for whom, for otherwise comparable journeys, train was not available. A high degree of variability within the sample could thus be expected.

For model evaluation, and also due to practical considerations, the sample was divided into two different sets of sub-samples. The one was a random division into two equal groups, based on whether the reference number of the household was an odd or an even number; these two samples were coded as SBB1 and SBB2 (SBB being an abbreviation of Son, Breugel and Best). The other division was geographic, the division being into Son and Breugel (SB) and Best (B). The total set of data was identified by the code SBB. This division was intended to provide some possibilities for evaluating both model stability, by comparing the two random sub-samples, and geographic stability. A plan to select an additional geographic sample, Geldrop and a sector of Eindhoven (GE), had to be abandoned because of the work involved in preparing the necessary level of service data.

The sample was limited to those travelling by one, or a combination, of the six valid modes, making one or two home–work–home chains in the day, with the work address being a fixed place of work within the VODAR area. If the address of the place of work was the same as the residential address the observation was rejected. If two or more persons in a given household were observed as simultaneously making identical home–work–home journeys then only one observation was utilised; in this case either the head of the household was chosen or, if he (or she) was not one of the group, a selection was made on the basis of age and sex. The first priority was given to age, the second to sex; if two or more persons were in the same age–sex group then the first person on the file was selected.

The number of valid single chains ultimately selected was 390, with an average of 3·14 valid alternative modes per observation.

4.4.2 *The home–shop destination- and mode-choice model*

As with the definition of the type of home–work–home trips to be considered in the model, practical considerations and the desire to avoid unnecessary complications at this stage in the development of a modelling technique dictated a simplification of the problem. The modes included in the model were restricted again to car (driver), bicycle, moped, bus, train and foot. Whilst it was initially planned to consider only home–shop–home journeys, this restriction was later relaxed to include some journeys containing additional trips with a purpose other than to shop but with a destination within the same shopping centre as the shopping trip. As with the home–work mode-choice models, the restriction of the geographic distribution of the residential addresses in the sample had a number of very definite advantages. In the case of the home–shop chains this related not only to the collection of the level of service data but also to the identification of the real alternative shopping centres and the collection of data on those centres. The definition of alternative centres for the SBB sample presented fewer difficulties than many of the other possible samples. Furthermore, SBB also had a good distribution of car ownership and incomes as well as the availability in Best of all six modes.

4.5 Data

The data required for the models can broadly be classified thus:

1 *Valid alternatives:* mode and, in the case of the shopping model, destination; all these were available from the home-interview survey.

2 *Socio-economic data:* the available socio-economic data for both the trip-maker and his family were limited to those collected in the home-interview survey.

3 *Level of service data:* level of service data per origin and destination pair were not readily available; they therefore had to be obtained especially for this study.

4 *Shopping centre attraction data:* employment data, classified to a four-digit system, were available from the Agglomeratie Eindhoven; no

other quantitative data could be readily obtained.

5 *Other data:* in view of the extensive use of bicycles, and the possible influence of weather on mode-choice, weather data were obtained for the survey days from the meteorological records of Eindhoven airport.

4.5.1 *The chosen alternative*

Tables 4.2 to 4.9 give a summary of the number of observations per sample subset for home–work–home single chains. Tables 4.10 to 4.12 give similar data over the total sample for double and single home–work–home chains as well as for shopping trips. For home–work chains these data are also given for persons with a driving licence from car owning households ('car available') and others ('no car available').

Table 4.2

Chosen mode, single home–work chains, SBB1

	Car available		No car available		Total	
	Persons	Per cent	Persons	Per cent	Persons	Per cent
Car	78	73	0	0	78	43
Moped	9	8	24	32	33	18
Bicycle	17	16	37	49	54	30
Bus	0	0	6	8	6	3
Train	1	1	5	7	6	3
Foot	2	2	3	4	5	3
Total	107	59	75	41	182	100

Table 4.3

Chosen mode, double home–work chains, SBB1

	Car available		No car available		Total	
	Persons	Per cent	Persons	Per cent	Persons	Per cent
Car	20	65	0	0	20	38
Moped	1	3	1	5	2	,4
Bicycle	10	32	16	76	26	50
Bus	0	0	0	0	0	0
Train	0	0	0	0	0	0
Foot	0	0	4	19	4	8
Total	31	60	21	40	52	100

Table 4.4

Chosen mode, single home–work chains, SBB2

	Car available		No car available		Total	
	Persons	Per cent	Persons	Per cent	Persons	Per cent
Car	78	68	0	0	78	38
Moped	10	9	36	38	46	22
Bicycle	15	13	39	42	54	26
Bus	5	4	10	11	15	7
Train	2	2	3	3	5	2
Foot	4	4	6	6	10	5
Total	114	55	94	45	208	100

Table 4.5

Chosen mode, double home–work chains, SBB2

	Car available		No car available		Total	
	Persons	Per cent	Persons	Per cent	Persons	Per cent
Car	16	67	0	0	16	30
Moped	0	0	9	31	9	17
Bicycle	5	21	13	45	18	34
Bus	0	0	0	0	0	0
Train	0	0	0	0	0	0
Foot	3	12	7	24	10	19
Total	24	45	29	55	53	100

Table 4.6

Chosen mode, single home–work chains, Best

	Car available		No car available		Total	
	Persons	Per cent	Persons	Per cent	Persons	Per cent
Car	71	60	0	0	71	29
Moped	15	13	40	33	55	23
Bicycle	21	18	54	44	75	31
Bus	4	3	12	10	16	7
Train	3	2	8	7	11	5
Foot	5	4	8	7	13	5
Total	119	49	122	51	241	100

Table 4.7

Chosen mode, double home–work chains, Best

	Car available		No car available		Total	
	Persons	Per cent	Persons	Per cent	Persons	Per cent
Car	17	52	0	0	17	24
Moped	1	3	9	24	10	14
Bicycle	13	39	20	54	33	47
Bus	0	0	0	0	0	0
Train	0	0	0	0	0	0
Foot	2	6	8	22	10	14
Total	33	47	37	53	70	100

Table 4.8

Chosen mode, single home–work chains, Son and Breugel

	Car available		No car available		Total	
	Persons	Per cent	Persons	Per cent	Persons	Per cent
Car	85	83	0	0	85	57
Moped	4	4	20	43	24	16
Bicycle	11	11	22	47	33	22
Bus	1	1	4	8	5	3
Foot	1	1	1	2	2	1
Total	102	68	47	32	149	100

Table 4.9

Chosen mode, double home–work chains, Son and Breugel

	Car available		No car available		Total	
	Persons	Per cent	Persons	Per cent	Persons	Per cent
Car	19	86	0	0	19	54
Moped	0	0	1	8	1	3
Bicycle	2	9	9	69	11	31
Bus	0	0	0	0	0	0
Foot	1	5	3	23	4	11
Total	22	63	13	37	35	100

Table 4.10

Chosen mode, single home–work chains, SBB

	Car available		No car available		Total	
	Persons	Per cent	Persons	Per cent	Persons	Per cent
Car	156	71	0	0	156	40
Moped	19	9	60	35	79	20
Bicycle	32	14	76	45	108	28
Bus	5	2	16	10	21	5
Train	3	1	8	5	11	3
Foot	6	3	9	5	15	4
Total	221	57	169	43	390	100

Table 4.11

Chosen mode, double home–work chains, SBB

	Car available		No car available		Total	
	Persons	Per cent	Persons	Per cent	Persons	Per cent
Car	36	66	0	0	36	34
Moped	1	2	10	20	11	11
Bicycle	15	27	29	58	44	42
Bus	0	0	0	0	0	0
Train	0	0	0	0	0	0
Foot	3	5	11	22	14	13
Total	55	52	50	48	105	100

Table 4.12

Chosen shopping-centre and mode, home–shop chains

(a) Best

	Shopping centre											Total
	1	2	3	4	5	6	7	8	9	10	11	
Car	6	2	2	7	0	0	0	0	0	5	0	22
Moped	0	2	0	2	0	0	3	0	0	4	0	11
Bicycle	0	7	12	32	2	0	0	2	0	56	0	111
Bus	12	0	0	0	0	0	0	0	0	0	0	12
Train	7	0	0	0	0	0	0	0	0	0	0	7
Foot	0	3	4	15	0	0	0	10	1	46	0	79
Total journeys	25	14	18	56	2	0	3	12	1	111	0	242

(b) Son and Breugel

	Shopping centre									Total
	1	12	13	14	15	16	17	18	19	
Car	18	3	0	1	0	0	0	0	37	59
Moped	0	0	0	0	0	0	0	0	11	11
Bicycle	1	1	0	4	1	0	0	0	73	80
Bus	2	0	0	0	0	0	0	0	0	2
Foot	0	3	0	4	0	0	0	0	31	38
Total journeys	21	7	0	9	1	0	0	0	152	190

4.5.2 *Definition of valid alternatives*

In order to minimise the volume of data that had to be handled, the valid alternatives for each person were identified. Thus the household and personal socio-economic data were examined to determine whether, on the basis of household vehicle ownership and possession of a driving licence, car, moped or bicycle could be regarded as alternative modes. If no vehicles were owned in any one of the three classes (car, moped, bicycle) then it was assumed that that mode was not a valid alternative. Similarly, if the person was not in possession of a valid driving licence for cars, even if a car was available within the household, then car was not considered to be a valid alternative.

If a destination was more than 2km away, walk was not considered a relevant alternative mode. Train was considered relevant only for trips from Best to Eindhoven. Bus was not considered a relevant alternative for work trips within Son and Breugel. Bus was also not considered a valid mode for shopping trips unless there was at least one stop between the boarding and alighting stops.

Alternative shopping centres were determined by tabulating the total number of shopping destinations per 100 square metres for Best, and Son and Breugel residents. After more detailed examination of a number of trip records, it was found that there were no valid (in the context of this study) shopping trips made by residents of Best to destinations outside Best but within the VODAR study area, except to the centre of Eindhoven. A similar result was found for Son and Breugel residents.

The individual shopping centres within Best, and Son and Breugel were determined by tabulating the retail employment data in a similar manner to the shopping trip destinations and then checking these results in the field. This led to the identification of ten centres in Best and eight in Son and Breugel (Figs 4.7 and 4.8). Since trips between shops within a shopping centre had not necessarily been reported in the survey, the centre of Eindhoven could be considered only as a single centre; in the context of this study, however, this was the most realistic approach, since it would be viewed by most shoppers as a single alternative.

4.5.3 *Socio-economic data*

Of the available socio-economic data, the following were utilised in the models estimated:

Income: household income, as collected in the study.

Car availability: cars per driver, i.e. the ratio of cars in a household to the number of persons with a licence.

Moped availability: the ratio of mopeds in a household to the number of persons aged 15 years and over (although 16 is the legal lower age limit for riding mopeds, the age classification system utilised in the survey prevented the use of the legal age).

Bicycle availability: the ratio of bicycles in a household to the number of persons aged five years and over.

Occupation: a broad classification of occupations into white and blue collar. A simple definition of white collar workers was applied: all workers in an occupation with a first digit of 1, 2 and 3, coded to the 1970 Dutch Central Bureau of Statistics system (CBS, 1971), were allocated to white collar, the rest to blue collar.

4.5.4 *Level of service data: home–work chains*

Level of service data were required for each of the valid modes pertaining to each home–work chain in the sample. In practice this meant obtaining data for up to five modes for each of the 390 observations.

With the exception of cost and terminal data, all level of service data for home–work chains were prepared by the study sponsor, Projectbureau Integrale Verkeers- en Vervoerstudies.

(a) *Network data.* Network data were derived by manually locating each pair of home and work addresses on large-scale plans. Travel times and distances for car, bicycle and moped were estimated using networks and travel timetables prepared for Agglomeratie Eindhoven by Buro Goudappel en Coffeng. The speeds in the car networks were based on a

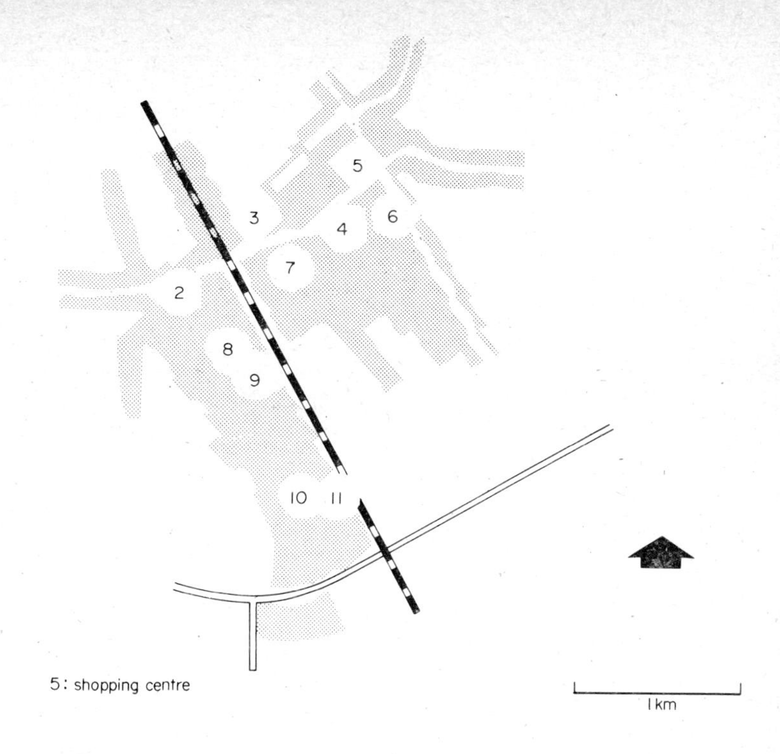

Fig. 4.7 Shopping centres, Best

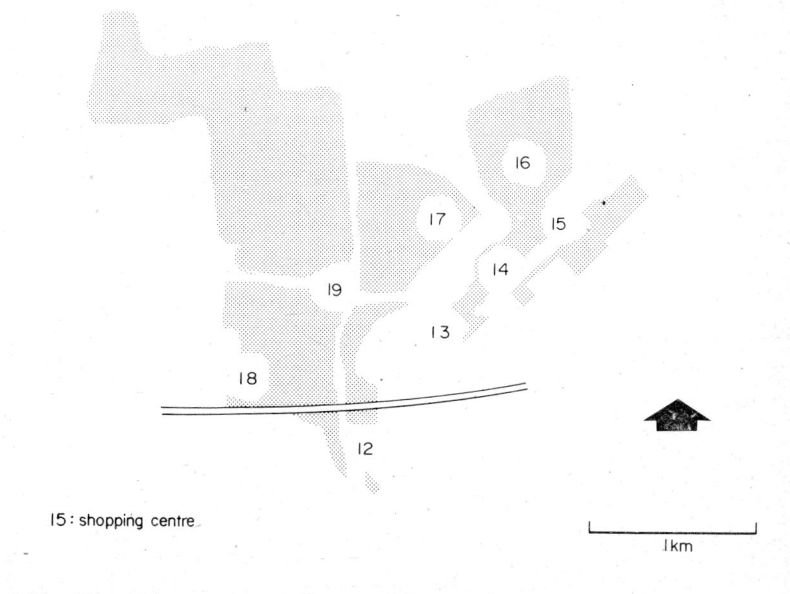

Fig. 4.8 Shopping centres, Son and Breugel

journey time study carried out by the Rijkswaterstaat in 1970.

The networks and matrices were based on a zoning system and therefore the individual values derived from the network had to be corrected to allow for the positive or negative difference between the location of the zone centroid and the individual address (home or work) for which data were being sought.

Since the type of car driven by car users, or available to those not driving, was not known, an average cost for a typical family car (Vauxhall Viva) was supplied by the ANWB, the Dutch motoring organisation; this was 15·4 cents per km for variable costs (this can be broken down as 5·0 cents per km for depreciation etc., 5·7 cents per km for fuel and 4·7 cents per km for maintenance, tyres etc.). However, in all the work done in the course of this study only the fuel costs were used, in keeping with traditional modal split cost measurement procedures.

An average speed of 12km per hour was assumed for bicycle and of 24km per hour for moped. It was assumed that bicycle costs were not significant; no costs were therefore attributed to bicycle usage. An average fuel cost of 2·0 cents per km was allocated to moped usage; this cost was based on data supplied by RAI, the Dutch trade association for both motor vehicles and bicycles.

All distances for walking trips were measured from large-scale plans. An average speed of 6km per hour was assumed, with a maximum permitted distance of 2km (20 minutes).

Journey times by bus and train were derived from the 1970 timetables. Bus was not considered a valid alternative for trips within Son and Breugel, since no such bus trips were observed, but it was assumed a valid mode for trips within Best.

The fares paid by those choosing public transport, and perceived by those for whom it was not chosen but was apparently a valid alternative, cannot be determined; the former because the type of ticket utilised was not known and the latter because the completeness of their knowledge was unknown. After examination of the 1970 costs, it was decided to use the values given in Table 4.13. For bus trips, 6- or 12-trip tickets proved cheaper than season tickets.

All walking distances between home or work and the relevant bus stop or station were measured from large-scale plans. Travel between Eindhoven station and work addresses more than 2km from the station was assumed to be by bus. Transfer times utilised were 7 and 15 minutes for bus to bus and 9 and 17 minutes for train to bus; the two-minute difference between bus to bus and train to bus being the result of the extra walking distance involved in making a train to bus transfer. The two

Table 4.13

Public transport fares, home–work chains (Dutch cents)

Trip type	Mode	Fare (return), cents	Ticket type
Within Eindhoven	Bus	83	6-trip ticket
	Train	–	–
Within Best	Bus	54	6-trip ticket
	Train	–	–
Within Son and Breugel	Bus	–	–
	Train	–	–
Best to Eindhoven	Bus	113	6-trip ticket
	Train	85	Monthly season ticket
Son and Breugel to Eindhoven	Bus	117	12-trip ticket
	Train	–	–

different values reflect the two frequencies of services within Eindhoven, 15 and 30 minutes (waiting time was assumed to be one half of the headway).

(b) *Terminal data.* Data available from the survey did not enable the specific location of the parking place, or potential parking place, for cars, bicycles or mopeds to be identified for each individual observation. After discussions with Philips and DAF and a number of measurements of 'typical' situations, the average values given in Table 4.14 were adopted for car, bicycle and moped.

As far as is known no charges were levied by employers in the Eindhoven area in 1970 for the use of private parking spaces. While there were some metered spaces in the centre of Eindhoven, there was also a large number of free parking spaces available. It was therefore decided to assume that no charge was made for parking private cars. Waiting time for both bus and train was assumed to be five minutes. No terminal times were associated with foot trips.

4.5.5 *Level of service data: home–shop chains*

For home–shop chains, level of service data were required for each of the possible alternative shopping centres from each individual home address in

Table 4.14

Terminal or excess times, home–work chains

Work address	Mode	Occupation	Time (minutes)				
			Departure from home	Arrival at work	Departure from work	Arrival home	Total
Philips and DAF	Bicycle/moped	All	1	2	2	1	6
	Car	White collar	1	2	2	1	6
	Car	Rest	1	4	4	1	10
Centre of Eindhoven	Bicycle/moped	All	1	2	2	1	6
	Car	White collar	1	2	2	1	6
	Car	Rest	1	5	3	1	10
Elsewhere	Bicycle/moped	All	1	1	1	1	4
	Car	All	1	2	2	1	6
All	Foot	All	0	0	0	0	0

the sample. The limitation of the valid alternative centres to those within the municipality of residence and the centre of Eindhoven (section 4.5.2) considerably simplified the collection of level of service data.

(a) *Network data.* For both Best, and Son and Breugel the complete street systems were coded in a network, with each sample address, shopping centre (including Eindhoven centre) and bus stop being treated as a node. Speed assumptions were the same as those used for the home–work chains, except for car. Since the Rijkswaterstaat journey-time survey did not cover the full range of highway types forming the network, and the study budget did not permit a complete survey, some typical routes were surveyed and speeds for each link estimated on the basis of these data. Journey times and distances by foot, bicycle, moped and car, as well as walking distances to and from bus stops, were then calculated.

Journey times for bus and train were calculated from the 1970 time-tables, with walking distances derived from the street network; for the larger shopping centers (Eindhoven) and the linear centres a centroid was estimated and distances measured to and from that. Costs for car and moped were assumed to be the same as those used for the home–work model. Costs associated with walking and bicycling were again assumed to be zero. Bus fares were all assumed to be the day-return rate given in Table 4.15. The return train fare between Best and Eindhoven was taken as the second-class day-return fare: 120 cents.

Table 4.15

Bus fares, home–shop chains (Dutch cents)

Trip type	Fare (return)
Within Best	70
Within Son and Breugel	Not applicable
Best to Eindhoven	155
Son and Breugel to Eindhoven	130

(b) *Terminal data.* The terminal data utilised for foot, bicycle, moped and car were derived from a number of observations (Table 4.16). For bus trips a waiting time of ten minutes was assumed and for train five minutes.

Table 4.16

Terminal or excess times, home–shop chains

Destination	Mode	Departure from home	Arrival at shopping centre	Departure from shopping centre	Arrival home	Total
Son, Breugel and Best	Bicycle/moped	1	1	1	1	4
	Car	1	2	2	1	6
Eindhoven centre	Bicycle/moped	1	1	1	1	4
	Car	1	6	6	1	14
All	Foot	0	0	0	0	0

In Best, and Son and Breugel there were no parking charges, and, whilst meters were in use in the centre of Eindhoven (10 cents per 30 minutes), there were many free spaces. It was therefore decided to assume that there was no parking cost associated with shopping trips. Although this may be an over-simplification, the estimation of the perceived costs for those who travelled by car and for those who chose not to do so was outside the scope of this study.

4.5.6 *Shopping centres*

At least one quantitative measure of the attractiveness, or relative attractiveness, of the shopping centres was required, and ideally more than one, so that some comparative evaluation of different measures could be carried out. Three basic characteristics initially considered were retail floor space, turnover, and employment. Consideration was also given to some sort of empirical, quantitative rating system. Of all these, the only information readily available for all the centres was employment. Floor space data could have been collected and an empirical rating system developed, but in view of the resources available for the complete study it was decided to limit the work to employment data. These were available coded to a four-digit coding system based on the industrial coding system of the Dutch Central Bureau of Statistics (CBS, 1970).

The employment data for each centre were extracted from the files of Agglomeratie Eindhoven and, in the case of Best and Son and Breugel, coded to the 10 metre co-ordinate reference. Employment data were also obtained for some service industries, such as banks, restaurants, barbers, and estate agents, factors which could affect the relative attractiveness of a shopping centre.

The survey data did not reveal whether anyone visited more than one shop in a shopping centre, nor was a differentiation made between visits to shops to purchase durable or consumer goods. The number of jobs in retail and in other services per shopping centre are given in Tables 4.17 and 4.18.

Table 4.17

Employment per shopping centre, Son and Breugel

Centre no.	Retail and related employment	Other service employment
12	22	0
13	1	0
14	26	0
15	2	0
16	10	0
17	4	0
18	4	0
19	206	81

4.5.7 *Weather data*

It was originally hypothesised that, in view of the lack of protection offered by both bicycle and moped against inclement weather, mode-choice could be affected by weather. Weather data were therefore obtained from the meteorological station at Eindhoven airport, some 7km south of Best, for each day to which the trip data related.

For work trips a simple weather variable was derived, according to

Table 4.18

Employment per shopping centre, Eindhoven and Best

Area	Centre no.	Retail and related employment	Other service employment
Eindhoven centre	1	3,724	2,804
Best	2	38	0
	3	72	85
	4	58	13
	5	51	26
	6	15	0
	7	6	11
	8	5	2
	9	3	0
	10	90	5
	11	17	1

whether or not it rained between 0600 and 0900 hours. For shopping trips a more complex coding system was derived, based on the period 0900 to 1800 hours.

A simple analysis of the observed modal split for home–work chains (Table 4.19) showed, however, that it was most unlikely that this variable could contribute anything useful to the model, and it was therefore never used.

Table 4.19

Weather and modal split, home–work chains

Weather	Mode (figures in percentages)						Total	Number of observations
	Car	Moped	Bicycle	Bus	Train	Foot		
Dry	37	16	31	4	1	11	100	380
Wet	35	17	38	2	2	6	100	266

5 The Work Mode-Choice Model

5.1 Introduction

The general modelling methodology and the multinomial logit model used in this study were described in Chapter 3, and the data available for this study as well as the general strategy of the empirical work were described in Chapter 4. The purpose of this chapter is to present the results of the models estimated for work trips.

Two basically different groups of models were estimated. One was a mode-choice model for travellers who make a single home–work–home chain during the day. The second group of models include travellers who return home during lunch time and therefore make a double home–work–home chain during the day.

Primary emphasis was placed on the single chain model because it represents a less complex behaviour; it is also the more common situation in larger metropolitan areas where distances to work tend to be greater than in the VODAR area.

5.2 The theoretical model

The work mode-choice model explains the conditional probabilities of choosing a mode of travel for the work trip given the residential and employment locations and given that a trip is made. Thus, the dependent variable can be denoted as follows:

$$P(m{:}M_{dt})$$

or

$$P_t(m|d)$$

where m denotes an alternative mode and M_{dt} denotes the set of available modes to destination d for traveller t.

The logit model predicting this probability is written as:

$$P(m{:}M_{dt}) = \frac{e^{U_{mt}}}{\sum_{m' \epsilon M_{dt}} e^{U_{m't}}} \tag{5.1}$$

where U_{mt} is the utility of mode m to traveller t for the work trip to destination d (see (3.23).

In general we write the utility of a given mode as:

$$\begin{aligned} U_{mt} &= X_{mt}'\theta \\ &= \sum_{k=1}^{K} X_{mtk}\theta_k \end{aligned} \tag{5.2}$$

where a single coefficient vector was denoted as θ for all modes (see 3.25). However, the utilities of the modes are different because the vector of variables X_{mt} assumes different values for different modes.

If a variable appears only in the utility function of mode m then it is a mode m specific variable which takes a value of zero in all other modal utilities. If a variable appears in the utility function of all modes, then it is a generic variable. The value of a generic variable must not be equal for each alternative (i.e. each mode) for all observations, or, mathematically, this variable is cancelled out (see section 3.5.2).

5.3 Variables used in the models

The explanatory variables used in the work mode-choice models are of two types; level of service characteristics, and socio-economic variables. We have denoted the variables symbolically by using abbreviations. For example, the variable 'in-vehicle travel time' is written as *IVTT*. If a variable is specified as mode specific then this is indicated by either prefixing or suffixing the following letters to the variable name:

walk	*W*	
car	*C*	
bicycle *(fiets)*	*F*	*TW* (two-wheel)
moped *(bromfiets)*	*BF*	
bus	*B*	*PT* (public transport)
train	*T*	

Thus if, for example, the variable '*IVTT* by car' appears in the utility function of car with a coefficient different to that of other modes then it is denoted as *CIVTT*.

The variable name *CON* denotes a constant.

5.3.1 *Level of service variables*

The level of service variables used in the models included in the report were:

IVTT – in-vehicle travel time (in minutes). For walking trips, *IVTT* is always zero; for all mechanical modes its the time spent in or on the vehicle.

OVTT – out-of-vehicle travel time (in minutes). For walking trips, *OVTT* is the total walking time of the trip. For car, bicycle and moped it is denoted as *POVTT*, which is defined as the time taken to walk to and from the parked vehicle, bicycle or moped, as well as to park and unpark. For bus and train, *OVTT* consists of two parts: *WSOVTT* and *SOVTT*. *WSOVTT* is defined as the time spent walking to and from the bus stop or station. *SOVTT* is the time spent at a bus stop or station, as well as in transferring from a bus to a train or vice versa.

OPTC – out-of-pocket travel cost (in Dutch cents). For walk and bicycle trips *OPTC* has a zero value. For car and moped trips it was assigned a value equal to the fuel costs, in keeping with traditional expectations of perceived motoring costs; no parking charges were included (see section 4.5). For bus and train *OPTC* equals the costs of the fares specified in section 4.5.

5.3.2 *Socio-economic variables*

The socio-economic variables used in the models included in this report were:

HHINC – annual household income. These data were available coded to the following classes: (1) less than Hfl.5,000 a year; (2) Hfl.5,001–10,000 a year; (3) Hfl.10,001–15,000 a year; (4) Hfl.15,001–20,000 a year; (5) Hfl.20,001–25,000 a year; (6) more than Hfl.25,000 a year.

PER – the number of persons in the household aged five years or older

AOD – the number of private cars and non-commercial vans reported to be in the possession of the household, divided by the number of licensed drivers in the household. *AOD* was not permitted to have a value in excess of one.

BOP – the number of bicycles reported as being owned by the household divided by the number of persons aged five years or older in that household.

MOA – the number of mopeds reported as being owned by a household, divided by the number of persons aged 15 years or over in that household.

HHPOS – position in the household. This variable equals one for the head of the household, otherwise it equals zero. Since the purpose of this variable in the mode-choice model was to represent car availability, it was also assigned a value of one in the case of adults with a driving licence who were not the head of the household, if there was perfect car availability – i.e. if the variable *AOD* for that household was equal to one.

OCC – the occupation of the traveller. This was used as a simple dummy variable taking the value of one for professionals, managers, and executives, and otherwise taking a value of zero.

A number of other variables available from the original data files, such as age and sex, were considered, as was a more detailed description of the variable *OCC*, but these were excluded during the course of the work on the basis of *a priori* considerations or simply due to the limited number of different specifications which could be estimated.

5.4 Alternative specifications

Various means of introducing the explanatory variables into the model were tried. Every model estimated was based on a different specification of the modal utility functions. Several initial specifications were formulated based on *a priori* reasoning and past experience, including the results obtained from the earlier models.

The initial estimation runs were deliberately restricted to rather simple specifications and were initially done with only half of the total sample (SBB1), since the level of service data for the full SBB sample were not available until later in the study.

The results of these initial runs indicated that out-of-pocket travel costs, *OPTC* (whether or not it was divided by income), had a small positive coefficient, which was not significantly different from zero. This means that – at least in this particular case – out-of-pocket travel costs do not significantly influence modal choice. This result corresponds with an *a priori* assumption made in a mode-choice study with data from Amsterdam and Rotterdam (de Donnea, 1971), in which it was assumed that out-of-pocket cost has no influence on the choice whether or not to drive a car to work in the existing Dutch urban transportation conditions. The results obtained from this study could, however, be explained by the small differences between the costs of the different modes due to the generally low level of transport costs and the relatively short length of the trips in the sample.

Previous mode-choice models have tended to introduce the level of service characteristics as generic variables; see for example PMM (1972), CRA (1972), and Ben-Akiva (1973). In this study this practice was not necessarily justified for the various components of out-of-vehicle travel time. This is one of the few studies ever conducted with totally disaggregate data – that is, data in which the home and work addresses could be precisely located – and thus the walking time to and from bus stops or stations could be accurately appraised. Parking time could, however, only be estimated, since no information was available on the location of the parking place; waiting and transfer times could also only be estimated. Thus, the different components of out-of-vehicle travel time were estimated with varying degrees of reliability; aggregation to give total out-of-vehicle travel time was therefore avoided in most of the specifications tried. The work by Stopher, Spear and Sucher (1974) on the measurement of inconvenience in urban travel suggests, however, that a division of out-of-vehicle travel time into its various components could be preferable to aggregation into a single variable. An exploratory run on the SBB1 sample using in-vehicle travel time *(IVTT)* as a mode specific variable indicated that *IVTT* could be reasonably applied as a generic variable, since little difference was found between the mode specific coefficients (see Table 5.1, in which, as in all other tables in this chapter and Chapter 6, the coefficients of the variables are shown in normal numerals and the standard errors in italics; the statistics given at the foot of each table are explained in section 3.5, with the exception of χ^2 which is equal to $-2(L^*0-L^*\theta)$, where $L^*\theta$ is the value of L^* for the vector of estimated coefficients and L^*0 is the value of L^* for $\theta=0$, where L^* denotes the value of the *ln* likelihood function; the degrees of freedom are indicated by d.f.). This suggests that the traveller values in-vehicle travel time the same, regardless of mode. From other results it was also found that logarithmic transformation of the travel time variables (*IVTT* or *OVTT)* had lower explanatory power than the untransformed forms.

Based on essentially the above considerations, the specification of the level of service variables which appeared to be the most suitable was:

IVTT as a generic variable;
WOVTT as a mode specific variable for walk;

Table 5.1

Coefficients of in-vehicle travel times estimated with SBB1 sample

Mode	Coefficients *Standard error*
Car	–0·0997 *0·0584*
Bicycle	–0·0995 *0·0258*
Moped	–0·1273 *0·0516*
Bus	–0·0759 *0·0297*
Train	–0·0881 *0·0580*

POVTT as a mode specific variable for car, bicycle and moped;
WSOVTT as a mode specific variable for bus and train;
SOVTT as a mode specific variable for bus and train.

The major differences between the models reported in this chapter, and summarised in Table 5.2, are in the specifications of the socio-economic variables and the modal constants. The inclusion of socio-economic variables in the utility function is clearly intended to explain differences in choice between different groups of persons. It can also facilitate the use of the model as a predictive model for different market segments; this is illustrated in the results of the aggregate forecasting tests discussed in section 5.7.

Modal constants have a totally different function from the other variables. If the variables included in the modal utility functions fully explain the mode-choice behaviour then the modal constants, or more generally the pure alternative effects, should equal zero. Thus, with a perfect model specification and with perfect data, it can be argued that no constants are necessary. Estimating a model without constants is in practice not recommended, however, since the estimated values of the coefficients of the variables included could be seriously affected if those variables do not fully explain the observed behaviour.

The constants, or pure alternative effects, represent therefore the effect of those variables which influence mode-choice but were not explicitly included in the model. As was explained in section 3.5.2, the formulation of the logit model is such that constants have to be alternative specific, or, in this particular case, mode specific; a constant cannot be generic. If we have reason to believe that those variables which should have been included in the model to make it complete, but were in fact excluded, have different values for different situations, then the values of the constants will also differ. Under such circumstances the use of a model estimated on data for one area, to predict behaviour in another area, or at a different period of time, or for a different socio-economic group, may well be questionable. In a well-specified mode-choice model the modal constants can be primarily attributed to level of service variables such as reliability, comfort, privacy, convenience, etc., many of which are either difficult or impossible to measure. An attempt was made in this study to account for the pure alternative effects through the introduction of various mode specific socio-economic and vehicle availability variables which could usually be expected to be highly correlated with a modal constant. The exclusion of constants was then considered acceptable if the coefficients of the various level of service variables were not significantly affected by omission of the constant. Thus, from Table 5.2 it can be seen, for example, that the replacement of *FCON* and *BFCON* by *FHHINC* and *BFHHINC* had no effect on the coefficient of *IVTT* (models 1 and 3 compared with model 5).

From the results of some of the early models estimated, it was observed that the variables *HHPOS* and *OCC* (position in household and occupation) had no significant influence on mode-choice so far as the sample of data used for this study was concerned.

A major problem can be observed from the results of the forecasting tests reported in section 5.7. Apparently both the level of service and socio-economic variables included in the various models do not adequately explain the choice of moped, although they seem to be extremely effective with regard to other modes. This can probably be explained by a lack of suitable variables describing safety, age, comfort or social background which could explain the differences in the usage of moped among various groups of travellers. If this were the case, then the constant, or the element of the utility function not explained by the other variables included, could be expected to vary with different combinations

Table 5.2
Estimation results of alternative specifications with SBB sample

Variable	Model number													
	1	2	3	4	5	6	7	8	9	10	11	12	13	14
IVTT	−·0673 *·0091*	−·0757 *·0106*	−·0672 *·0116*		−·0644 *·0101*	−·0732 *·0101*	−·0664 *·0107*	−·0649 *·0105*	−·0643 *·0102*	−·0644 *·0103*	−·0665 *·0094*	−·0715 *·0090*	−·0721 *·0092*	−·0600 *·0093*
lnIVTT			−·0343 *·5016*	−1·5639 *·3971*										
OVTT	−·1193 *·0172*		−·1419 *·0346*	−·1610 *·0234*										
lnOVTT			·5186 *·7366*											
WOVTT		−·1676 *·0291*			−·1661 *·0321*	−·1545 *·0293*	−·2535 *·0535*	−·2527 *·0535*	−·2511 *·0530*	−·2201 *·0527*	−·1257 *·0266*	−·1333 *·0266*	−·1095 *·0281*	−·1192 *·0295*
POVTT					−·2933 *·0794*	−·2372 *·0727*	−·2314 *·1018*	−·2255 *·1013*	−·2145 *·0907*	−·2539 *·0920*	−·3559 *·0786*	−·3590 *·0795*	−·2675 *·0919*	−·3260 *·0946*
COVTT		−·3290 *·0803*												
FOVTT		−·2501 *·1015*												
BFOVTT		−·5141 *·1040*												
WSOVTT		−·1206 *·0210*			−·1195 *·0218*	−·1149 *·0209*	−·1131 *·0222*	−·1113 *·0220*	−·1101 *·0213*	−·1041 *·0214*	−·0987 *·0193*	−·1034 *·0193*	−·1269 *·0238*	−·1136 *·0234*
SOVTT		−·0674 *·0236*			−·0962 *·0344*	−·0887 *·0329*	−·0836 *·0341*	−·0834 *·0341*	−·0828 *·0340*	−·0936 *·0345*	−·0853 *·0306*	−·0511 *·0216*	−·0824 *·0289*	−·0856 *·0288*
OPTC/HHINC	·0066 *·0101*		·0086 *·0115*	·0384 *·0100*	·0081 *·0107*	·0119 *·0105*	·0067 *·0113*	·0071 *·0113*	·0071 *·0113*	·0210 *·0119*	·0189 *·0113*			
CHHINC							−·5690 *·2616*	−·5764 *·2616*	−·5899 *·2559*	−·5866 *·2567*				
CAOD	·5877 *·4396*	1·1363 *·0528*					·7647 *·5048*	·7721 *·5034*	·7636 *·5024*	1·5314 *·5444*	2·1651 *·4753*	2·2928 *·4709*	2·2025 *·4690*	
*CAOD*HHINC*			·2408 *·1120*	·4968 *·1162*	·2435 *·1418*									
*CAOD*HHINC/PER*						·2963 *·4183*								
*CAOD*lnIVTT*														1·0056 *·1800*
COCC								−·1213* *·4928*						
FCON	·3254 *·3514*		·7486 *·6009*	1·4463 *·5795*										

FHHINC					·0286 *·1138*		−·5294 *·2422*	−·5332 *·2421*	−·5393 *·2408*	−·6860 *·2448*				
FHHINC/PER						·2480 *·3527*								
FBOP										1·5790 *·3841*	1·4698 *·3399*	1·3272 *·3263*	1·4936 *·3410*	1·4348 *·3211*
FOCC							·3037 *·5665*							
BFCON	−·7845 *·3517*		−·3864 *·4712*	·3563 *·3673*										
BFHHINC					−·2562 *·1081*		−·8109 *·2411*	·8209 *·2409*	−·8236 *·2406*	−·8091 *·2530*				
BFHHINC/PER						−1·1043 *·3626*								
BFMOA										1·8492 *·7457*	2·2659 *·5558*	−·0132 *·5375*	·2281 *·5527*	·6689 *·5536*
BFOCC							−·1390 *·6262*	·0197 *·4537*						
TWOCC								·0497* *·4537*						
PTCON	1·8615 *·5976*		1·8984 *·8316*	3·1672 *·7681*									1·7404 *·9155*	1·5057 *·9252*
PTHHINC					·1099 *·1823*		−·4557 *·2707*	−·4594 *·2696*	−·4553 *·2690*					
PTHHINC/PER						·1670 *·4585*								
PTOCC							·1654 *·6759*	·1086* *·5897*						
$L^*(0)$	−435·13	−435·13	−435·13	−435·13	−435·13	−435·13	−435·13	−435·13	−435·13	−435·13	−435·13	−435·13	−435·13	−435·13
$L^*(\hat{\theta})$	−272·81	−271·69	−271·17	−292·60	−269·14	−272·08	−266·55	−266·83	−266·86	−257·59	−268·69	−270·10	−268·29	−260·54
χ^2	324·63	326·88	327·92	285·05	331·97	326·10	337·16	336·59	366·53	355·07	332·88	330·04	333·68	349·17
d.f.	7	8	9	7	10	10	14	13	11	12	9	8	9	9
ρ^2	·37	·38	·38	·33	·38	·37	·39	·39	·39	·41	·38	·38	·38	·40
$\bar{\rho}^2$	·37	·37	·37	·32	·37	·37	·38	·38	·38	·40	·38	·37	·38	·40

* Three different versions of model 8 were estimated, each one incorporating only one of the three variables *COCC*, *TWOCC*, and *PTOCC*; all the coefficients, except for those of *TWOCC* and *PTOCC*, and statistics given are those for the model in which *COCC* was included.

of those variables and thus a simple constant would not adequately cover the missing variables. For this reason the constants and the socio-economic variables relating to the two two-wheel modes, bicycle and moped, were kept separate in spite of the fact that at first sight these two modes may appear to have similar unobserved attributes.

The constants and the socio-economic variables introduced in the public transport utility were combined for bus and train, primarily because of the small number of trips by both of these two modes.

In general all the variables described above were introduced into the modal utility functions in a linear form. Only in a few cases was an attempt made to explore non-linear finite functions of the variables; this was done with the time variables and also with the car availability variable *(AOD)*, which was utilised only as a car specific variable *(CAOD)*. For the time variables a natural log transformation was tried. For the *CAOD* variable two finite functions were tried; the one was a product of *CAOD* and household income *(HHINC)* and the other was the product of *CAOD* and the natural log of in-vehicle travel time by car: *CAOD lnIVTT*.

In a model with a maximum of six alternatives, only five (alternative specific) constants, or coefficients of a given socio-economic variable, can be identified (see section 3.5). The walk mode was therefore used as the base alternative and the coefficients of mode specific variables, such as income, should be interpreted relative to the walk mode.

The estimation results for the alternative specifications tried with the full SBB sample (390 trips) are given in Table 5.2. Models 5 to 14 are based on the final specification selected for the level of service variables, whilst models 1 to 4 show some of the other alternative specifications tried. The differences among models 5 to 14 are in the specifications of the various alternative specific constants as well as the socio-economic and vehicle availability variables. The estimation results and the differences between the models are discussed in detail in the following sections.

5.5 Discussion of the estimated coefficients

5.5.1 *Values of the estimated coefficients*

The strongest *a priori* knowledge which we have about the estimated values of the coefficients is with regard to their signs. We expect that, with everything else held equal, a deterioration in the level of service offered by any mode will reduce the probability of that mode being chosen. Thus an essential requirement is that the utility of any one mode should decrease as the value of most level of service variables increases (this is not the case, of course, with a level of service variable such as comfort, if comfort was measured on a scale which increased with increasing comfort). If a given level of service variable enters a utility function only once, then, with the exception of some specific transformations, it can be expected that the coefficient of that variable will be negative. If, however, the variable is entered in more than one form, e.g., as a simple variable and in a logarithmic transformation, then it is possible that only one of the coefficients need be negative. Thus in model 3 (Table 5.2), for instance, the sum of ($-0{\cdot}14\ OVTT + 0{\cdot}52\ lnOVTT$) decreases with increasing values of *OVTT*, if $OVTT > 3{\cdot}5$ minutes. (*OVTT* was always $> 3{\cdot}5$ minutes in the data-set used for estimation); in fact in this particular case the coefficient of *lnOVTT* was not significantly different from zero. This general requirement is satisfied in all the models estimated with the exception of *OPTC* in every model in which it was included. Since the coefficient of *OPTC* was never found to be significantly different from zero, out-of-pocket travel costs were assumed not to have any significant influence on mode-choice in this particular sample and they were therefore ultimately excluded. A similar result was obtained with the joint destination- and mode-choice shopping model, as reported in Chapter 6.

There are also some *a priori* expectations with respect to the relationships between certain coefficients of level of service attributes. For example, one would expect the coefficient of an out-of-vehicle travel time variable to be greater than the coefficient of in-vehicle travel time. In all the specifications tried *IVTT* was, indeed, found to have a lower coefficient than any of the *OVTT* variables; the coefficients of walk as a mode *(WOVTT)* and walk to a bus stop or station *(WSOVTT)* are almost equal and have approximately twice the value of the coefficient for in-vehicle travel time. Walking thus appears to be twice as inconvenient as riding in, or on, a vehicle. This relationship corresponds to the usual assumption used to create generalised costs in Wilson type models in the UK, where the coefficient of excess time is usually taken to be twice that of in-vehicle travel time (McIntosh and Quarmby, 1970). Waiting time at a bus stop or station appears to be more inconvenient than in-vehicle travel time, but not so burdensome as walking. This is a departure from the usual assumption mentioned above, as well as from US studies in which the coefficient of waiting time is usually taken to be 2·5 times that of in-vehicle time (e.g., Pratt and Deen, 1967). However, the relatively low coefficient of waiting time in this sample might be attributable to a highly reliable transport service, and therefore to an over-estimate of the value of waiting time at the station and bus stops. Furthermore, it could also reflect other errors in estimating the waiting times utilised in this study. It should also be noted that, compared with the coefficients of other level of service variables, the coefficient of *SOVTT* has a relatively large variance. This is probably due to the low variability of headways in the public transport services available to people in the sample (i.e. low variability in the values of *SOVTT*). Since the coefficient of *OPTC* was never found to be statistically significant, it is not possible to draw any conclusions on the absolute value of time. Had a coefficient significantly different from zero been estimated for *OPTC*, values of time could have been estimated by dividing the coefficients of the time variables by that of *OPTC*.

The expectations with respect to the values of both the constants and the coefficients of the socio-economic variables are more complicated and rely on very limited, or non-existent, past experience. One particular problem encountered in the design of the study was the limited availability of results from previous studies utilising similar models, especially for European circumstances.

Everything being equal, one would expect that as car ownership increases the probability of choosing car as the mode of transport would increase and, thus, that the probability of choosing other modes would decrease. One would therefore expect that the coefficient of car availability *(CAOD)* would be positive, and, for similar reasons, that the coefficients of both bicycle and moped availability would also be positive.

Household income and modal constants appear in the utility function of more than one mode and therefore interpretation must relate to their relative values and not their absolute values. In model 5, for example, the relative values of the coefficients of household income indicate that as household income increases, and everything else is held constant, the increase in the probability of using car is relatively greater than that for other modes, while the probability of choosing a moped will decrease relative to all other modes. Between these extremes are public transport, which decreases relative to car but increases with respect to the other modes, bicycle, which decreases relative to car and public transport and increases relative to walk and moped, and walk, which is a base mode. Thus the probability of walking increases relative to moped but decreases relative to all other modes; this could reflect the socio-economic status of a moped as a transport mode.

The modal constants and socio-economic variables could be interpreted as representing the pure preferences for the alternative modes if the utility derived from the level of service characteristics was equal across all modes. A direct interpretation of this kind is easier when all the level of service characteristics are introduced as generic variables. For example, these variables in model 1 imply that if *IVTT* and *OVTT* are equal across modes then someone with perfect car availability, i.e. with *AOD* equal to one, will rank the modes in the following order: public transport, car, bicycle, walk, and, last, moped. The place of public transport in this pure ranking appears to be too high. However, the pure ranking for a person with all modes perfectly available given by model 13 (which represents an improved specification) is: car, public transport, bicycle, moped, and, lastly, walk. This preference ordering agrees more closely with *a priori* expectations, and is also implied by model 14, which was the last specification estimated with this sample, and which we consider to have the most satisfactory specification of the models estimated.

5.5.2 *The stability and statistical reliability of the estimated coefficients*

The reliability and stability of the estimated coefficients can be observed in several ways, including the relative magnitudes of the standard errors, and the variability of the estimates both across different specifications and across different sub-samples.

The magnitude of the standard errors of the estimated coefficients (compared with the magnitude of the estimated coefficients) is relatively small for the travel time variables, but is comparatively higher for some of the modal constants and socio-economic variables – in particular, the moped specific variables. This could also be observed from the variability of the estimated coefficients of the same variables across different specifications.

From Table 5.2 it is evident that the coefficients of travel times are quite stable, in particular the coefficient of *IVTT*. On the other hand, some of the coefficients of the socio-economic variables and the constants appear to be less stable. This pattern was also observed in the comparison of different sub-samples.

As recorded in section 4.4, two types of sub-samples were created from the SBB sample. One was a random division into SBB1 and SBB2; a division which proved of value in the evaluation of the statistical stability of the estimated coefficients in comparison with the estimated values of the standard errors. The second was a division into two selected geographic sub-samples, SB and B. This allowed a comparison of coefficients between two different areas, with differences between coefficients from the two areas being used to trace possible specification errors. We assumed that travel behaviour in both areas is similar and that

Table 5.3

Estimation results, model 5, for five data sets

Variable	Data set				
	SBB	SBB1	SBB2	B	SB
PTHHINC	·1099	·3866	−·0633	−·0955	−·0528
	·1823	*·3205*	*·2362*	*·2274*	*·6001*
FHHINC	·0286	·3703	−·2722	−·2247	·2830
	·1138	*·1777*	*·1678*	*·1628*	*·1980*
BFHHINC	−·2562	·0265	−·4835	−·4710	−·0351
	·1081	*·1711*	*·1576*	*·1563*	*·1775*
*CAOD*HHINC*	·2435	·5316	·0739	−·0685	·6278
	·1418	*·2341*	*·1945*	*·1860*	*·2766*
IVTT	−·0644	−·0739	−·0573	−·0689	−·0501
	·0101	*·0155*	*·0140*	*·0117*	*·0265*
WOVTT	−·1661	−·1365	−·2028	−·2075	−·2046
	·0321	*·0526*	*·0447*	*·0426*	*·1322*
POVTT	−·2933	−·2356	−·3675	−·3418	−·2382
	·0794	*·1225*	*·1140*	*·0969*	*·1597*
WSOVTT	−·1195	−·0887	−·1591	−·1333	−·1102
	·0218	*·0306*	*·0333*	*·0272*	*·0519*
SOVTT	−·0962	−·1803	−·3968	−·0704	−·1047
	·0344	*·0650*	*·0361*	*·0380*	*·1200*
OPTC/HHINC	·0081	·0278	−·0062	−·0071	·0335
	·0107	*·0175*	*·0137*	*·0144*	*·0268*
$L^*(0)$	−432·13	−201·98	−233·15	−282·25	−152·88
$L^*(\hat{\theta})$	−269·14	−109·53	−149·72	−186·92	−73·09
χ^2 (d.f. = 10)	331·97	184·89	166·86	190·65	159·57
ρ^2	·38	·46	·36	·34	·52
$\bar{\rho}^2$	·37	·44	·34	·33	·50
No. of observations	390	182	208	241	149

Table 5.4

Estimation results, model 13, for four data sets

Variable	Data set			
	SBB	SBB1	SBB2	B
IVTT	−·0721	−·0836	−·0664	−·0769
	·0092	*·0146*	*·0123*	*·0110*
WOVTT	−·1095	−·1074	−·1148	−·1098
	·0281	*·0494*	*·0349*	*·0313*
POVTT	−·2675	−·3038	−·2699	−·2732
	·0919	*·1459*	*·1247*	*·1106*
WSOVTT	−·1269	−·0849	−·1776	−·1374
	·0238	*·0331*	*·0378*	*·0284*
SOVTT	−·0824	−·1121	−·0585	−·0788
	·0289	*·0518*	*·0339*	*·0307*
CAOD	2·2025	3·4796	1·5391	1·4234
	·4690	*·8296*	*·6052*	*·5504*
FBOP	1·4936	2·3249	·9527	1·3537
	·3410	*·5569*	*·4518*	*·3989*
BFMOA	·2281	1·3823	−·5096	·1808
	·5527	*·8503*	*·7479*	*·6329*
PTCON	1·7404	1·2698	2·2079	1·7068
	·9155	*1·5845*	*1·1724*	*·9961*
$L^*(0)$	−435·13	−201·98	−233·15	−282·25
$L^*(\hat{\theta})$	−268·29	−106·26	−153·41	−188·96
χ^2 (d.f. = 9)	333·68	191·45	159·47	186·58
ρ^2	·38	·47	·34	·33
$\bar{\rho}^2$	·38	·46	·33	·32
No. of observations	390	182	208	241

Table 5.5

Estimation results, model 14, for five data sets

Variable	Data set				
	SBB	SBB1	SBB2	B	SB
IVTT	–·0600	–·0679	–·0556	–·0688	–·0338
	·0093	*·0148*	*·0127*	*·0112*	*·0208*
WOVTT	–·1192	–·1210	–·1253	–·1164	–·2381
	·0295	*·0532*	*·0366*	*·0324*	*·1621*
POVTT	–·3260	–·3753	–·3464	–·3241	–·3785
	·0946	*·1580*	*·1279*	*·1130*	*·1939*
WSOVTT	–·1136	–·0701	–·1668	–·1269	–·0881
	·0234	*·0317*	*·0378*	*·0280*	*·0549*
SOVTT	–·0856	–·1187	–·0583	–·0828	–·0594
	·0288	*·0511*	*·0335*	*·0308*	*·0856*
CAOD lnIVTT*	1·0056	1·5746	·7718	·7254	1·6427
	·1800	*·3570*	*·2221*	*·2145*	*·4336*
FBOP	1·4348	2·2351	·9502	1·3529	1·4551
	·3211	*·5483*	*·4213*	*·3724*	*·6911*
BFMOA	·6689	1·8694	–·0314	·5000	1·3234
	·5536	*·8818*	*·7496*	*·6267*	*1·4021*
PTCON	1·5057	1·0927	1·8657	1·5553	·4079
	·9252	*1·5822*	*1·1873*	*1·0075*	*2·6117*
$L^*(0)$	–435·13	–201·98	–233·15	–282·25	–152·87
$L^*(\hat{\theta})$	–260·54	–101·43	–149·92	–186·02	–68·29
χ^2 (d.f. = 9)	349·17	201·11	166·44	192·47	169·18
ρ^2	·40	·50	·36	·34	·55
$\bar{\rho}^2$	·40	·49	·34	·33	·54
No. of observations	390	182	208	241	149

therefore a specification which has similar coefficients for both areas is superior to a specification with divergent estimates.

Tables 5.3, 5.4 and 5.5 give the estimation results for models 5, 13 and 14 for the different sub-samples. From these it can be seen that the coefficients estimated for model 14 are quite stable between the SBB and B samples. The variability between the estimates of the coefficients for SBB1 and SBB2 is considerably higher, but this can probably be attributed to the fact that the standard errors are larger as a result of smaller sample sizes.

This pattern of decreasing standard errors with increasing sample size for the three models and the sub-samples, given in Tables 5.3, 5.4 and 5.5, can be observed from Figs 5.1, 5.2 and 5.3. These figures indicate that an increase in sample size beyond 300 observations does not significantly reduce the standard errors of the coefficients of most of the variables. This pattern suggests that for these models a desirable sample size would be between 300 and 400 observations.

In order to compare the stability of the estimated coefficients between two independent random samples, a total sample of at least 600 observations would therefore be necessary, rather than the 400 observations available. Samples larger than 300–400 observations may, however, be necessary if more or other socio-economic variables were to be included. With the existing sample these have been found to have very large standard errors, in contrast to coefficients of *LOS* variables for which reasonable levels of reliability at smaller sample sizes were achieved.

Geographical comparison of the estimated coefficients between B and SB is not conclusive, because of the small number of observations in the SB sample. However, the stability between SBB and B seems satisfactory in view of the fact that, although B is included in SBB, there are still significant differences between both the means and distributions of several of the explanatory variables.

5.6 Analysis of alternative specifications

Of all the model specifications tried, that for model 14 appears to be the most satisfactory. The reasoning leading up to this conclusion, as well as aspects of some of the other models estimated, are discussed in this section.

As discussed in section 5.4, the formulation of the level of service variables in models 5 to 14 seems to be superior to that utilised in models 1 to 4. The coefficients of all the level of service variables incorporated in

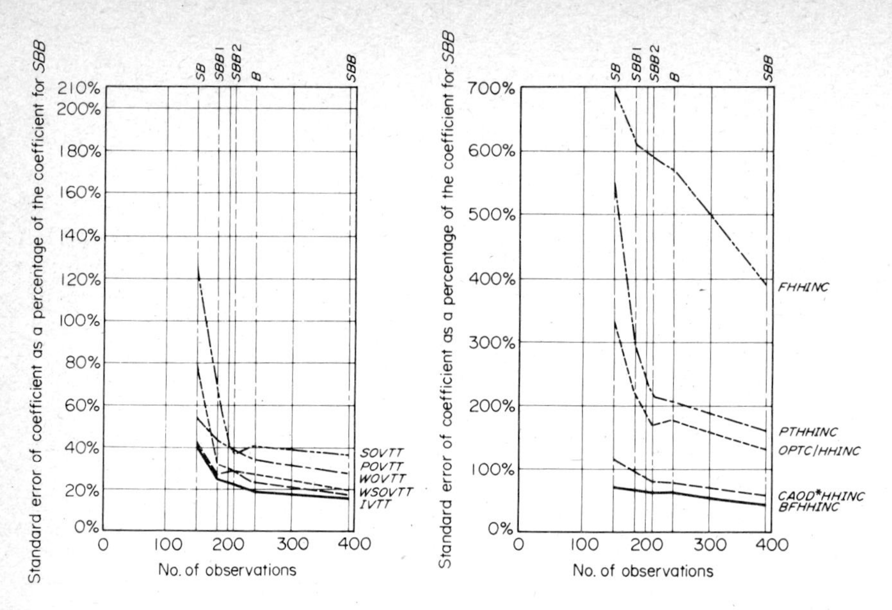

Fig. 5.1 The standard error of estimated coefficients and sample size, model 5

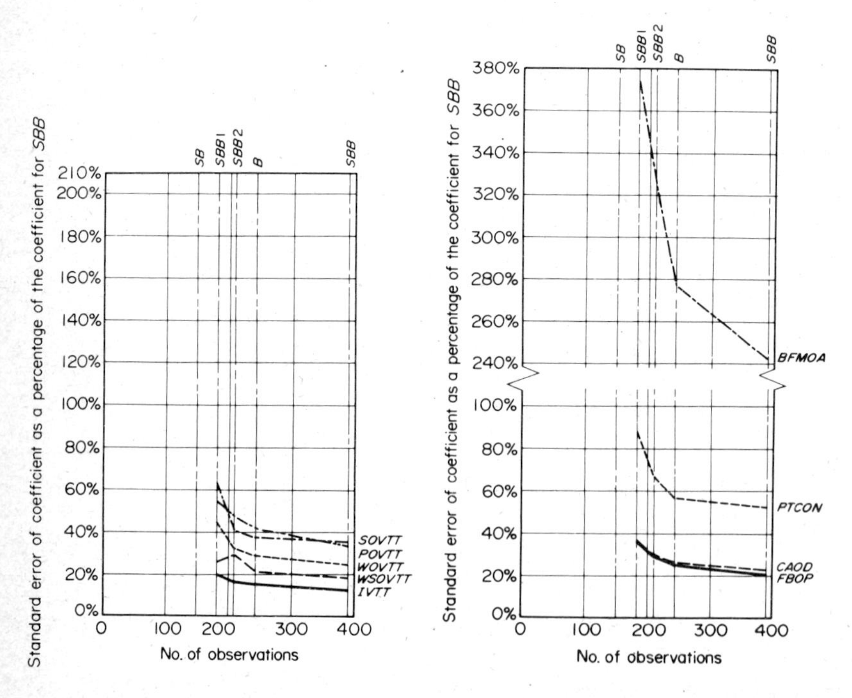

Fig. 5.2 The standard error of estimated coefficients and sample size, model 13

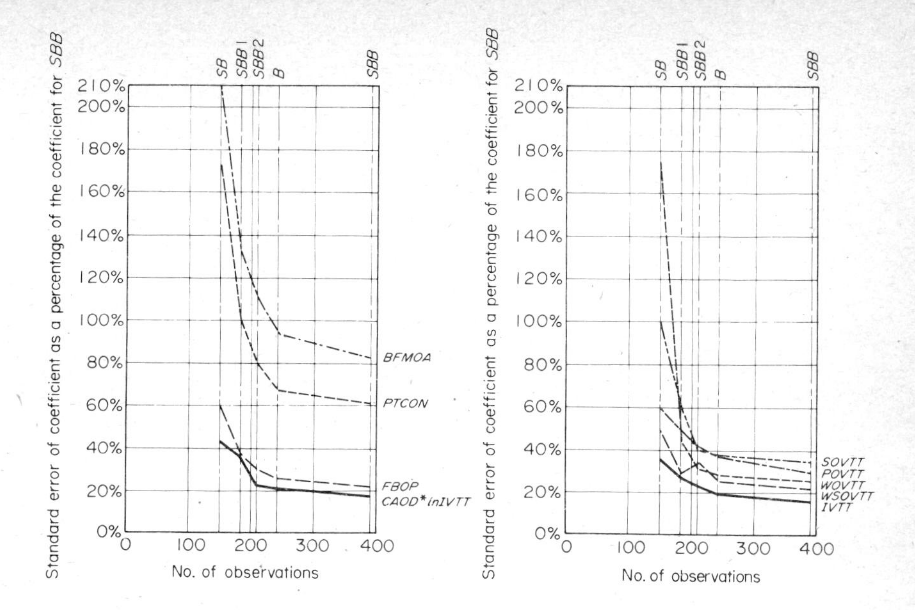

Fig. 5.3 The standard error of estimated coefficients and sample size, model 14

models 5 to 14 have the expected signs (with the exception of *OPTC/HHINC*, which is not significant) as well as the expected relative values. Therefore any preference between specifications 5 to 14 must be based on the behaviour of the socio-economic variables and the modal constants.

Models 5 and 6 have identical specifications, with the exception of household income, which was included in model 5 but replaced in model 6 by income per person. It would seem reasonable that the pure modal preferences would be more closely related to the total household income than to the average income per person, because of the effects of income pooling. Although model 5 has a slightly better goodness of fit than model 6 (see section 3.5.6), the estimation results are by no means conclusive evidence that model 5 is, indeed, any better than model 6. However, taken together with the previous statement, model 5 was considered superior to model 6.

In models 7, 8 and 9 the car specific variable *COAD**HHINC* was split into two separate variables, *CAOD* and *CHHINC*. In addition an attempt was made in models 7 and 8 to introduce occupation *(OCC)* as a variable (in the form of a mode specific variable), but in neither model were any of its coefficients significantly different from zero. Model 9 was considered to be less satisfactory than model 5 because of the relatively larger variance of the coefficient of *CAOD;* this is probably attributable to a high level of collinearity between car availability and household income.

The specification of model 10 includes the same variables as model 9 with, in addition, bicycle and moped availability *(BFOP* and *BFMOA).* In model 10, however, the variances of the coefficients of the socio-economic variables were large, and it thus appeared desirable to select only a subset of these variables for the following models. This was done in models 11, 12 and 13, in which only the vehicle availability variables were included. These variables were selected since it seemed reasonable to assume that they have a greater direct bearing on mode choice than household income. Furthermore, the coefficients of these variables in earlier models were more significant than those of the income variables.

Models 11 and 12 are identical, with the exception of *OPTC/HHINC,* which was excluded from model 12. In model 13 a public transport constant *(PTCON)* was reintroduced. Since it is highly probable that the specification of model 12 was not perfect, and, therefore, that the absence of a public transport constant could considerably affect the values of the coefficients of other variables, model 13 was considered the best of the series of models 7 to 13.

So far, then, models 1 to 13 have been evaluated and models 5 and 13 have been selected as two of the best models. These two models represent two essentially alternative specifications of the socio-economic variables. Model 5 is based upon mode specific income variables, whereas model 13 is based upon mode specific vehicle availability variables. These two models were also estimated for the various sub-samples and the results of these estimations are given in Tables 5.3 and 5.4. If we compare these two tables and also Figs 5.1 and 5.2, it is evident that model 13 is more stable than model 5. From a consideration of both goodness of fit and the significance of the coefficients it can be concluded that model 13 is also superior to model 5.

However, examination of the variability of the estimated coefficients of model 13 between the different sub-samples, given in Table 5.4, shows the coefficient of *CAOD* to be particularly unsatisfactory. Examination of the characteristics of the various sub-samples shows that the coefficient of *CAOD* has a smaller value in those sub-samples having a shorter average-trip length and a larger value in sub-samples with a longer average-trip length. It therefore seems reasonable to assume that when a car is perfectly available to a person, then the longer the trip the more likely he is to choose the car. When a car is not perfectly available, and there is some degree of competition among different users of a car within a household, it would also seem reasonable to expect that those making longer trips will tend to have priority in the use of the car over those making shorter trips. Thus it seems reasonable to specify the coefficient of car availability as a function of trip length; this theory is also discussed in section 6.5. If this is so, then it could be expected that, within the context of the particular data set being used, this function will show a diminishing marginal effect with increasing trip length; therefore, the *CAOD* variable was multiplied by the natural log of in-vehicle travel time by car. This change was implemented in model 14.

A comparison of model 13 with model 14, from Tables 5.4 and 5.5 as well as from Figs 5.2 and 5.3, shows that model 14 is superior to model 13 in terms of stability of the coefficients, in the significance of the coefficients and of the goodness of fit. If all the models in Table 5.2 are compared in terms of their goodness of fit, it is interesting to note that although model 10 has a goodness of fit equal to model 14, model 14 has a smaller set of coefficients, and thus must be considered as being the better of the two according to this measure; no other model has a goodness of fit comparable to that of model 14.

To summarise, model 14 appears to have the most satisfactory specification of those models estimated and presented in Table 5.2. In subsequent sections of this chapter this model is tested for its forecasting abilities, and its coefficients are discussed in terms of the implied elasticities.

5.7 Prediction tests

Two types of prediction tests were applied: the first was a disaggregate test designed primarily to determine how well the model fits the observed data; and the second, with aggregate data, was designed to test the applicability of the model for aggregate predictions. Whilst in fact the model is used to calculate a 1970 situation, and thus could hardly be regarded as a prediction in the sense 'to foretell', it is felt that it best describes the spirit of the work. The use of the word 'prediction', as opposed to 'forecasting', conforms with the definitions given by de Neufville and Stafford (1971).

5.7.1 *Disaggregate predictions*

In disaggregate predictions the explanatory variables are used to predict individual mode-choice probabilities. These individual probabilities are summed across a group of travellers and compared with the observed modal shares for the same set of people. When the group of travellers consists of the complete set used in the estimation of the model, this test can be viewed as a test of goodness of fit. However, the estimation procedure utilised in this study guarantees that if a model specification includes a constant the results of such a test will be perfect for the alternative to which that constant relates (the proof of this is given in section 3.5.6). For model 14 this implies that a disaggregate prediction test with the complete data-set used in model estimation will reproduce perfectly the total public transport share, since a public transport constant was included *(PTCON)*. However, since no constant was used for any of the remaining four modes, this test is still meaningful for the split between the two public transport modes on the one hand, and the four other modes on the other. The results of this test are given in the first row of Table 5.6, and, as stated above, it can be seen that the split between public transport and the other four modes is indeed perfectly reproduced. The results between the individual modes within these two sets are, however, also extremely satisfactory.

Table 5.6

Disaggregate prediction results

Sample			Walk	Car	Bicycle	Moped	Bus	Train	Total
SBB	P	No.	10·73	156·04	107·64	83·58	22·07	9·91	390
		%	2·75	40·01	27·60	21·43	5·66	2·54	100
	O	No.	15	156	108	79	21	11	390
		%	3·85	40·00	27·69	20·26	5·38	2·82	100
SBB1	P	No.	5·24	73·71	53·02	34·76	10·10	5·19	182
		%	2·88	40·50	29·13	19·10	5·55	2·85	100
	O	No.	5	78	54	33	6	6	182
		%	2·75	42·86	29·67	18·13	3·30	3·30	100
SBB2	P	No.	5·51	82·31	54·64	48·84	11·98	4·72	208
		%	2·65	39·57	26·27	23·48	5·76	2·27	100
	O	No.	10	78	54	46	15	5	208
		%	4·81	37·50	25·96	22·12	7·21	2·40	100
B	P	No.	7·83	76·95	75·41	55·41	15·50	9·91	241
		%	3·03	31·93	31·29	22·89	6·43	4·11	100
	O	No.	13	71	75	55	16	11	241
		%	5·39	29·46	31·12	22·82	6·64	4·56	100

Table 5.6 continued

Sample			Walk	Car	Bicycle	Moped	Bus	Train	Total
SB	P	No.	2·91	79·07	32·24	28·19	6·59	0	149
		%	1·95	53·07	21·64	18·92	4·42	0	100
	O	No.	2	85	33	24	5	0	149
		%	1·34	57·05	22·15	16·11	3·36	0	100
Zone 100, 103, 100 to Eindhoven centre	P	No.	0	16·62	2·58	5·21	5·56	7·03	37
		%	0	44·92	6·96	14·09	15·03	18·99	100
	O	No.	0	18	1	5	6	7	37
		%	0	48·65	2·70	13·51	16·22	18·92	100
Zone 100, 103, 110 to Eindhoven elsewhere	P	No.	0	33·49	11·47	18·44	7·99	2·61	74
		%	0	45·26	15·50	24·92	10·80	3·52	100
	O	No.	0	28	13	21	8	4	74
		%	0	37·84	17·57	28·38	10·81	5·41	100
Zone 200, 210 to Eindhoven centre	P	No.	0	11·65	2·35	2·71	2·30	0	19
		%	0	61·30	12·35	14·25	12·11	0	100
	O	No.	0	13	1	3	2	0	19
		%	0	68·48	5·26	15·79	10·53	0	100
Zone 200, 210 to Eindhoven elsewhere	P	No.	0	30·92	7·70	10·02	3·35	0	52
		%	0	59·47	14·81	19·26	6·45	0	100
	O	No.	0	33	10	7	2	0	52
		%	0	63·46	19·23	13·41	3·85	0	100
Zone 100, total	P	No.	2·25	46·25	19·06	13·05	5·40	0	86
		%	2·61	53·78	22·16	15·17	6·28	0	100
	O	No.	2	51	19	10	4	0	86
		%	2·33	59·30	22·09	11·63	4·65	0	100
Zone 210, total	P	No.	·37	11·37	6·01	5·00	·25	0	23
		%	1·61	49·43	26·15	21·72	1·09	0	100
	O	No.	0	13	4	6	0	0	23
		%	0	56·52	17·39	26·09	0	0	100

P = Predicted, O = Observed

To provide a thorough test of the model one would ideally apply it to a second set of data, not used in the model estimation. Unfortunately because of the budget and time constraints pertaining to this study, a second set of data was not available. Instead, the test was applied to several subsets of the data set used for model estimation; the results of these are also given in Table 5.6. The subsets used include the sub-samples

SBB1, SBB2, SB and B as well as two other types of sub-samples which were created. The first of these consisted of groups of people with a home address in a specific zone, regardless of their work address, and the second was formed of groups with a home address in a specific zone (see Figs 5.4 and 5.5) and a work address either in the centre of Eindhoven or elsewhere in Eindhoven.

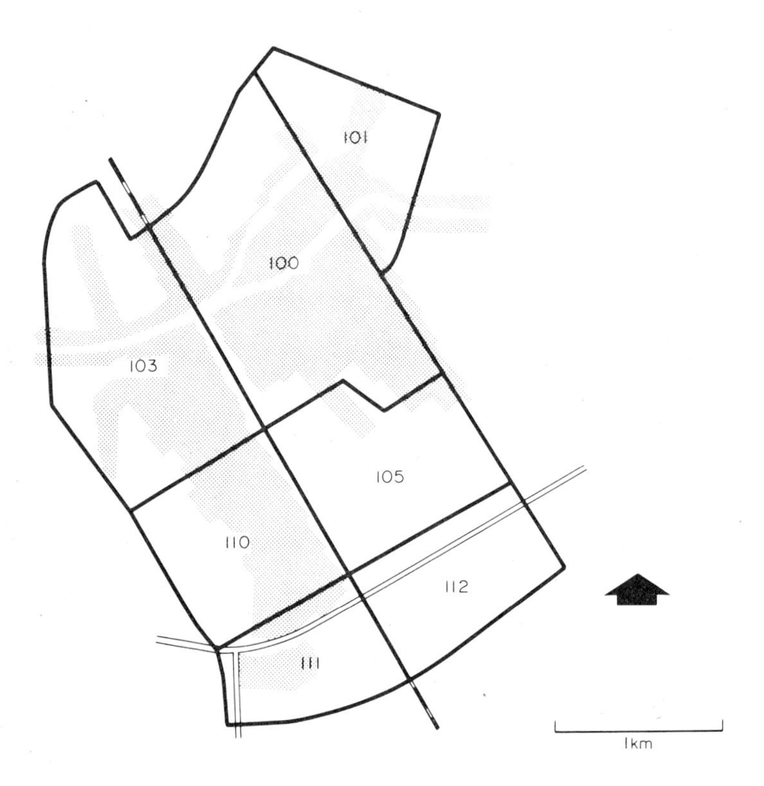

Fig. 5.4 Zoning plan, Best

The predicted modal shares for all these various subsets compare favourably with the observed shares (Table 5.6). The differences between the observed shares and the predicted shares are minimal for the larger subsets, i.e. those consisting of more than 100 observations. In the case of the smaller subsets, the differences between the observed and predicted shares are relatively large for bicycle and moped. There is, however, a tendency for some mutual compensation here, in that when one share is over-estimated the other is usually under-estimated; this means that the total share of bicycle and moped together is more satisfactorily reproduced than the individual shares.

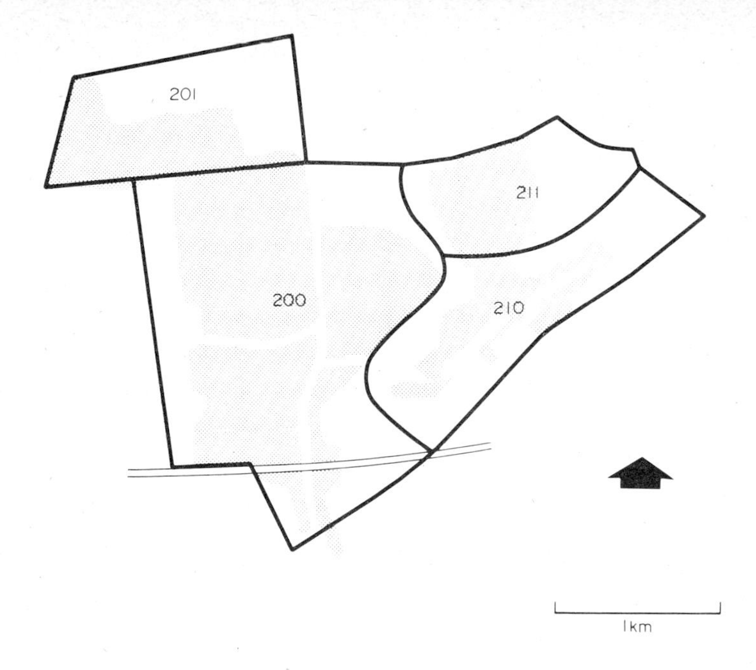

Fig. 5.5 Zoning plan, Son and Breugel

5.7.2 *Aggregate predictions*

Disaggregate prediction gives some indication as to how well the model fits the actual observed disaggregate data. However, in normal predictive work, disaggregate data are not usually available and thus the model must be applied as an aggregate model for predictions. As explained in section 3.6, simple substitution of group averages for the explanatory variables will result in a biased forecast of the average probability, or share; this bias will disappear only if all people in the group for which predictions are being prepared are identical in terms of the values of all the explanatory variables. Given the usual heterogeneity of urban populations as well as of level of service characteristics, this implies a very low level of aggregation.

Between the two extremes of disaggregate predictions and the use of averages for the entire group in aggregate predictions, it is possible that a stratification scheme, or system of market segmentation, could be identified in such a way that the aggregation bias can be assumed to be small and, within the context of the application, negligible.

Table 5.7 gives the results of aggregate predictions prepared using model 14 for four origin and destination pairs and using a very coarse zoning scheme. For this application of the model, all the travellers making

Table 5.7

Aggregate predictions, model 14: no market segmentation

Sample			Car	Bicycle	Moped	Bus	Train	Total
Best	P	No.	30·75	3·88	9·37	1·28	·73	46
to Eindhoven		%	66·85	8·43	20·37	2·78	1·59	100
Zone 2	O	No.	19	12	14	0	1	46
		%	41·30	26·09	30·43	0	2·17	100
Best	P	No.	24·14	1·79	7·13	3·59	3·38	40
to Eindhoven		%	60·35	4·48	17·83	8·98	8·45	100
zone 3	O	No.	18	1	6	7	8	40
		%	45·00	2·50	15·00	17·50	20·00	100
Son and Breugel	P	No.	23·69	3·31	9·33	·66	0	37
to Eindhoven		%	64·03	8·95	25·22	1·78	0	100
zone 2	O	No.	23	7	5	2	0	37
		%	62·16	18·92	13·51	5·41	0	100
Son and Breugel	P	No.	17·74	1·68	5·04	·54	0	25
to Eindhoven		%	70·96	6·72	20·16	2·16	0	100
zone 3	O	No.	16	2	5	2	0	25
		%	64·00	8·00	20·00	8·0	0	100

P = Predicted
O = Observed

a trip between any pair of zones were grouped together and the average value of each variable for the members of that group was utilised. The zoning plan for Eindhoven is given in Fig. 5.6. The calculated shares are clearly not very satisfactory in comparison with the observed shares. In particular, the share of car trips is significantly over-estimated for the first two groups, which consist of people residing in Best. Whilst the car share forecasts for the Son and Breugel groups are better, they are not satisfactory. The divergence between the observed and predicted shares serves to demonstrate the errors which can occur as a result of simply taking averages for each individual variable and directly applying them to the model.

In Son and Breugel the majority of the residents are car owners, whereas in Best there are many families without a car; the model therefore evidently performed better in cases of higher levels of car ownership than

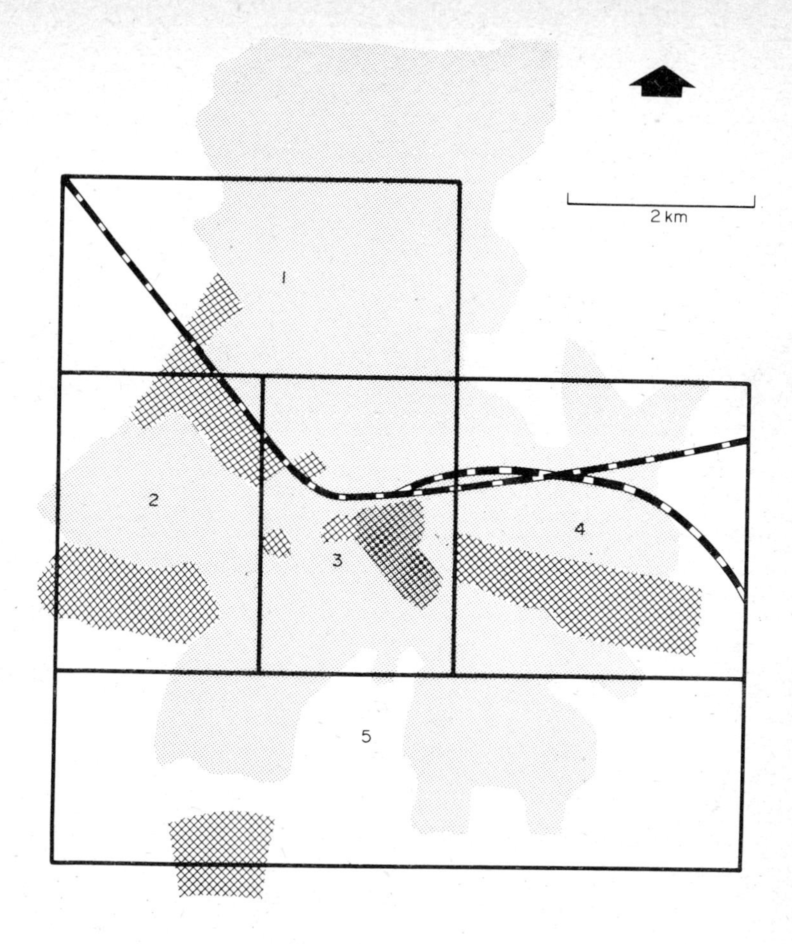

Fig. 5.6 Zoning plan, Eindhoven

lower. Through the use of one set of average values of all variables, including *CAOD*, applied to all travellers, a car has effectively been made available to people for whom in fact it was not available.

As discussed in section 3.6, the forecasting error can be reduced by a stratification of the travellers according to one or more characteristics which can be identified in the model specification. Of course, the ultimate stratification is that of complete disaggregation, but this would also produce some prediction errors, the magnitude of which depends upon the validity of the model itself. The errors in the predictions given in Tables 5.7 are therefore a combination of both disaggregate prediction errors and an aggregation bias. Given a specific model, it is possible to reduce the aggregate prediction error by attempting to reduce the aggregation bias.

In Table 5.8 the results of aggregate predictions are given for travellers between the same sets of specific zone pairs as used for the work summarised in Table 5.7 but now stratified into car owners and non-car owners. This sort of stratification is quite common in travel demand forecasting (see, for example, Wilson, 1969) and it is evident from the analysis of the differences in the errors of the predicted shares of car between Son and Breugel, and Best described above that such a stratification would improve the aggregate predictions.

Table 5.8

Aggregate predictions, model 14: market segmentation by car availability

Sample			Car	Bicycle	Moped	Bus	Train	Total
Car available (CAOD > 0)								
Best to Eindhoven zone 2	P	No.	24·50	1·34	3·33	·52	·31	30
		%	81·67	4·47	11·10	1·73	1·03	100
	O	No.	19	4	7	0	0	30
		%	63·33	13·33	23·33	0	0	100
Best to Eindhoven zone 3	P	No.	20·57	·89	4·52	1·69	1·33	29
		%	70·93	3·07	15·59	5·83	4·59	100
	O	No.	18	1	2	5	3	29
		%	62·07	3·45	6·90	17·24	10·34	100
Son and Breugel to Eindhoven zone 2	P	No.	22·23	1·80	5·55	·32	0	30
		%	74·10	6·00	18·50	1·07	0	100
	O	No.	23	6	1	0	0	30
		%	76·67	20·00	3·33	0	0	100
Son and Breugel to Eindhoven zone 3	P	No.	16·22	·99	3·51	·27	0	21
		%	77·24	4·71	16·71	1·29	0	100
	O	No.	16	1	3	1	0	21
		%	76·19	4·76	14·29	4·76	0	100
Sum over four groups	P	No.	83·52	5·02	16·91	2·80	1·64	110
		%	75·93	4·56	15·37	2·55	1·49	100
	O	No.	76	12	13	6	3	110
		%	69·09	10·91	11·82	5·46	2·73	110

Table 5.8 continued

Sample			Car	Bicycle	Moped	Bus	Train	Total
No car available (*CAOD* = 0)								
Best	P	No.	0	3·63	11·01	·89	·46	16
to Eindhoven		%	0	22·69	68·81	5·56	2·88	100
zone 2	O	No.	0	8	7	0	1	16
		%	0	50·00	43·75	0	6·25	100
Best	P	No.	0	1·32	3·89	2·32	3·47	11
to Eindhoven		%	0	12·00	35·36	21·09	31·54	100
zone 3	O	No.	0	0	4	2	5	11
		%	0	0	36·36	18·18	45·45	100
	P	No.	0	4·95	14·90	3·21	3·93	27
Sum over		%	0	18·33	55·19	11·89	14·56	100
two groups	O	No.	0	8	11	2	6	27
		%	0	29·63	40·74	7·41	22·22	100

P = Predicted
O = Observed

The results given in Table 5.8 can be considered acceptable in view of the zoning system adopted and the consequential effects of this on the values of the *LOS* variables applied. Furthermore, for each individual group there is also probably an error in the observed share, relative to the total population, as a result of sampling. Since this error increases with decreasing sample size, the greater the number of the observed travellers utilised to compute the observed aggregate share, the more meaningful the aggregate prediction tests. Thus an adequate evaluation of the aggregate prediction errors can be undertaken only for those zonal pairs with a high trip density, i.e. those in which the error in the observed share could be assumed to be negligble. Unfortunately, none of the individual groups of observations nor the total was sufficiently large to satisfy this criterion.

The disaggregate prediction tests have demonstrated that model 14 reproduces the average choice of people satisfactorily. The aggregate predictions have much more important implications with respect to the usefulness of the models, since the model is tested in the same way as that in which it will be generally applied. The aggregate mode-choice predictions, based on a stratification of travellers into those with a car available and those with no car available, can be considered as being satisfactory, given the limitations of the data used. This implies that the model can be usefully applied to aggregate predictions in transportation planning studies.

5.8 Elasticities

The variables in the mode-choice model which are of primary interest to a transportation planner are the level of service variables. In addition to the use of a model for conventional area-wide predictions it can also be used to give indications of the likely effects of changes in selected level of service variables, given that all other variables remain constant. Such analyses provide useful information for both the development and general appraisal of possible new policies.

The sensitivity of travellers to changes in the level of service variables is expressed in terms of elasticities. Two types of elasticity measures can be calculated. The first of these are point elasticities, which are valid for small changes and represent the trend at a given situation. The second type of elasticity is that of arc elasticity; an arc elasticity is valid for large changes and represents the effect of moving from one situation to another. Care must be taken in the interpretation of arc elasticities, if the selected arc implies an extrapolation beyond the range of values for which the model was estimated.

The sensitivity of travellers to changes in the level of service variables is expressed in terms of elasticities in order to avoid dependency on the measurement units. The mathematical derivation of elasticities in the logit model is discussed in sections 3.5.4 and 3.5.5.

5.8.1 *Point elasticities*

In the terminology of this particular mode-choice model, direct elasticity can be defined as the percentage change in the probability of choosing a mode resultant on a change in one of the level of service characteristics of that mode. A cross elasticity refers to the effect on the choice probability of a given mode due to a change in one of the level of service characteristics of another mode.

The direct elasticities in the logit model can be specified thus:

$$E^{P(m)}_{X_{km}} = \left[1 - P(m)\right] \theta_k \,.\, X_{km} \qquad (5.3)$$

where X_{km} is the level of service variable k of mode m.

The cross elasticities due to a change in X_{km} are:

$$E^{P(m')}_{X_{km}} = -P(m) \;\; \theta_k \;\; X_{km} \qquad (m' \neq m) \tag{5.4}$$

The cross elasticities of any mode m' $(m' \neq m)$ with respect to a change in a level of service variable of mode m are equal (see section 3.5.3).

The elasticities are linear with the choice probability. Therefore for a given value of the variable X_{km} it is possible to represent the elasticities as a straight line. Figs 5.7 and 5.8 give the direct and cross elasticities of *IVTT;* Figs 5.9 and 5.10 give the elasticities for *POVTT;* and Figs 5.11 and 5.12 give the elasticities for *WSOVTT* (all for model 14). Thus if, for example, the in-vehicle travel time *(IVTT)* by car is 30 minutes and the current choice probability of car is 0·7, then its direct elasticity is −0·54 and the cross elasticities of all other modes with respect to *IVTT* by car are +1·26.

In other words, given that the current car-choice probability is 0·7 and that *IVTT* for car is 30 minutes, an increase of 1 per cent in the value of car *IVTT,* i.e. 0·3 minutes, will result in a 0·54 per cent decrease in the car-choice probability, from 0·700 to 0·696, and a 1·26 per cent increase in all other mode-choice probabilities, from 0·300 to 0·304. The elasticities of a mode-choice probability for any of the three level of service variables, *IVTT, POVTT* and *WSOVTT,* can be read directly from this set of six figures.

A set of point elasticities for a specific group of travellers is given in Table 5.9; the group of travellers were car owners living in Best and working in zone 2 in Eindhoven (Fig. 5.6). The observed share for a given mode was used to determine the choice probability and the observed averages of the relevant set of variables were used as the X_{km}. The most significant, or maybe important, result given in this table is the elasticity of the bus mode-choice probability with respect to *WSOVTT;* this was calculated as being −6·72. This very large elasticity indicates that an increase in the number of bus stops, and therefore a reduction of walking distances, would be a highly effective way of implementing a policy of attracting more travellers to bus.

5.8.2 *Arc elasticities*

By definition, point elasticities are valid only for small changes in the value of one of the independent variables; they can thus indicate a trend only at a particular point. In order to evaluate the effect of large changes in any of the independent variables – that is, in any of the variables

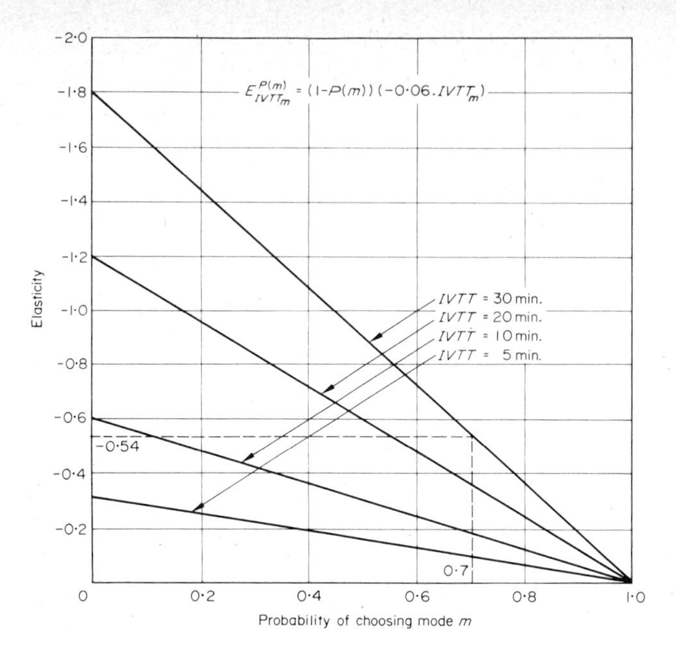

Fig. 5.7 Direct elasticity of in-vehicle travel time *(IVTT)*

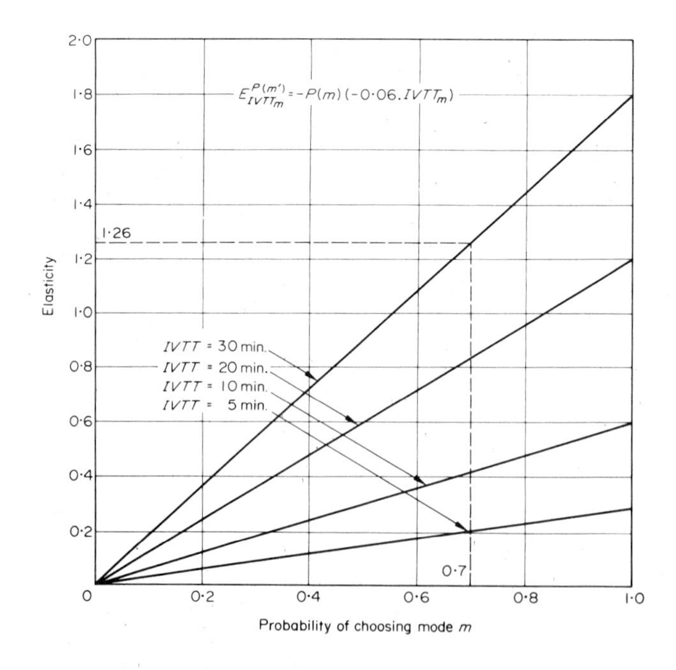

Fig. 5.8 Cross elasticity of in-vehicle travel time *(IVTT)*

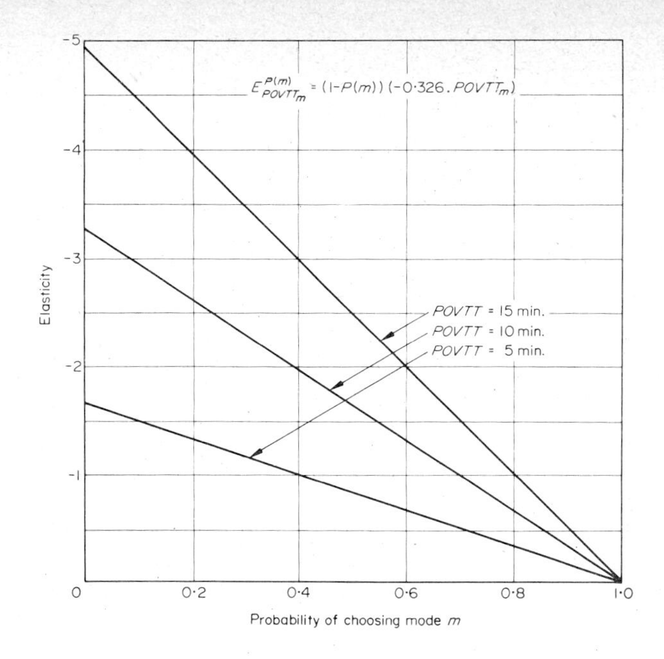

Fig. 5.9 Direct elasticity of park out-of-vehicle travel time *(POVTT)*

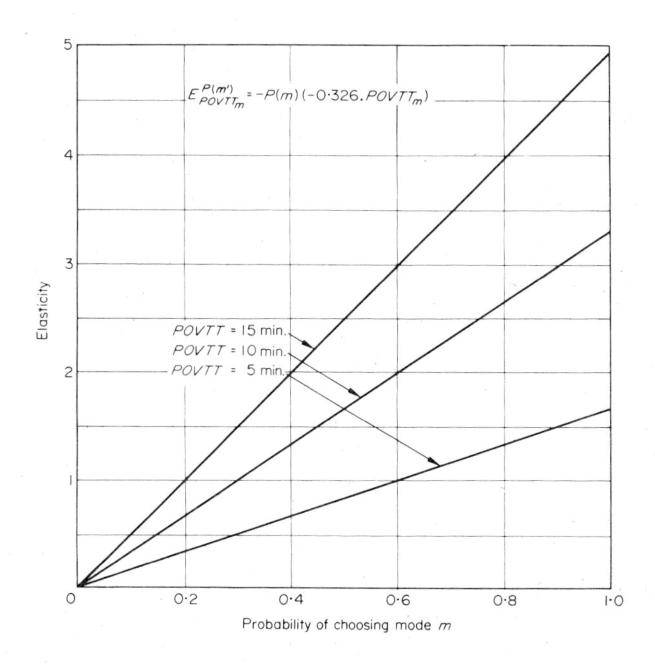

Fig. 5.10 Cross elasticity of park out-of-vehicle travel time *(POVTT)*

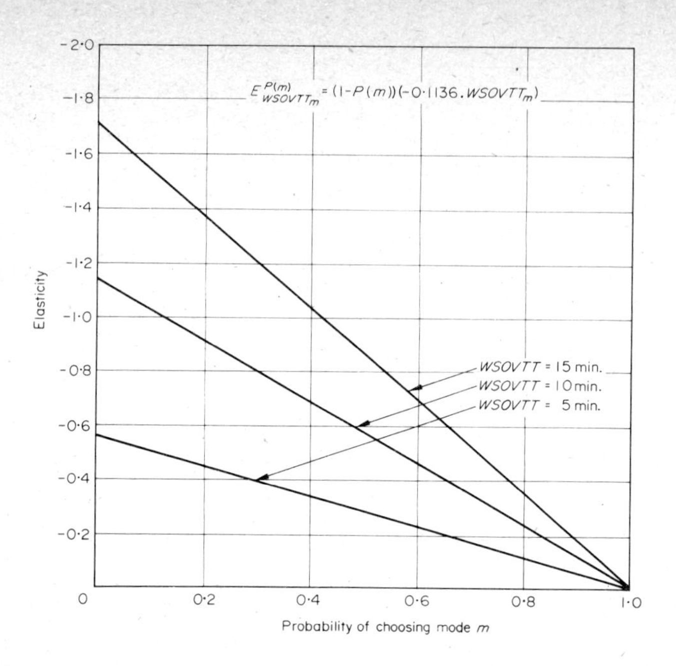

Fig. 5.11 Direct elasticity of walk to station out-of-vehicle travel time *(WSOVTT)*

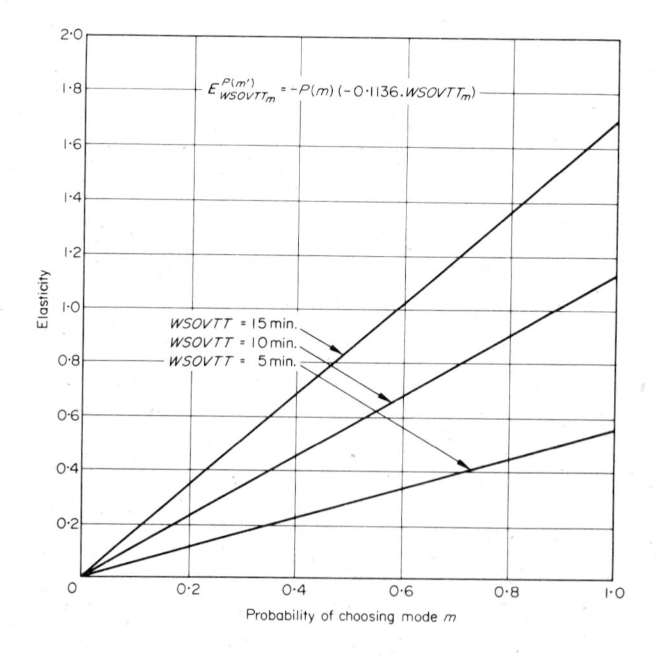

Fig. 5.12 Cross elasticity of walk to station out-of-vehicle travel time *(WSOVTT)*

Table 5.9

Direct point elasticities for trips between Best and Eindhoven, zone 2

Mode	Variable	Choice probability	Variable value	Elasticity
Car	*IVTT*	·817	24·0	−0·26
	POVTT	·817	8·0	−0·48
Bicycle	*IVTT*	·045	66·9	−3·83
	POVTT	·045	5·8	−1·81
Moped	*IVTT*	·111	34·5	−1·84
	POVTT	·111	5·8	−0·31
Bus	*IVTT*	·017	52·7	−3·11
	WSOVTT	·017	21·0	−6·72
Train	*IVTT*	·010	30·4	−0·16
	WSOVTT	·010	30·8	−0·90

incorporated in the utility function – the mode-choice probabilities have to be recalculated. A set of results of such a series of calculations is given in Table 5.10. These results apply to the same group of travellers as those given in Table 5.9. The first row of Table 5.10 gives the existing mode-choice probabilities. Thus, expressed in terms of percentages, 5·2 per cent choose bicycle; 17·1 per cent, moped; 77·0 per cent, car; and 0·7 per cent, public transport. The remaining rows of the table give the mode-choice probabilities resulting from large changes in a single variable, as well as from a combination of simultaneous, large changes in several variables.

Two examples of the application of this sort of analysis, which can be highlighted from Table 5.10, are the fact that a 10 per cent reduction of public-transport in-vehicle travel time has practically no effect on the mode-choice probability of car, whereas a 30 per cent increase in the parking out-of-vehicle travel time for car *(POVTT)* reduces the mode-choice probability of car to 60 per cent; associated with this are increases in the mode-choice probabilities of all other modes, including an increase in the probability of using public transport from 0·7 per cent to 1·2 per cent. From the last four rows of the table it can be seen that a combination of a number of measures can be extremely effective in changing the shares of the various transport modes. Thus, for example, if an objective was to reduce the level of car usage, a simultaneous 30 per cent increase in the time or walking distance to or from a parking place, a 20 per cent decrease in walking distance to or from a bus stop or station, as well as a 20 per cent decrease in waiting time, would reduce the probability of car usage by nearly 25 per cent (to 59 per cent), whilst increasing the probability of public transport usage from 0·7 to 3·2 per cent.

Table 5.10

Arc elasticities for trips by car owners ($CAOD \neq O$) between zones 100, 103, 110 and Eindhoven (except the centre)

Policy	Car *IVTT* (per cent)	Public transport *IVTT* (per cent)	Car *POVTT* (per cent)	Public transport *WSOVTT* (per cent)	Public transport SOVTT (per cent)	P (bicycle)	P (moped)	P (car)	P (public transport)
1						0·0517	0·1710	0·7702	0·0071
2	−10					0·0454	0·1503	0·7981	0·0061
3	+10					0·0586	0·1938	0·7397	0·0079
4		−10				0·0516	0·1706	0·7683	0·0095
5			+30			0·0886	0·2931	0·6063	0·0120
6		−10	+30			0·0882	0·2918	0·6037	0·0163
7			+30	−10	−20	0·0876	0·2897	0·5994	0·0233
8			+30	−20	−20	0·0868	0·2872	0·5941	0·0319
9		−10	+30	+20	+20	0·0891	0·2947	0·6097	0·0063

The differences between point and arc elasticities could be highly significant. For example, the first row of Table 5.9 gives a direct elasticity of *IVTT* by car of −0·26, whilst the arc elasticity for the same variable implied by the second row of Table 5.10 is −0·36.

5.9 Double chains

The home–work–home mode-choice models described so far in this chapter were based on people who reported only a single round trip from home to their place of work and home again during a 24-hour period. It is, however, relatively common in Dutch towns outside the major conurbations for people to return home from work for lunch. It is evident that the decision on whether or not to return home for lunch depends, among other things, on the distance between home and work or, more generally, on the level of service offered by the transport system for this trip, and thus the ease with which the round work–home–work trip can be made within the allotted lunch break.

One would also expect that the choice of mode for the journey to and from work would be dependent upon the individual's choice of whether to go home at lunchtime or not. For example, it is conceivable that under some conditions a car is the only feasible mode for a double chain, because of the location of the work place and the duration of the lunch break. Travel time by public transport may be too long to enable a trip home to be made during the lunch break; indeed, within the data-set utilised for this study no double chain was observed in which public transport was the chosen mode. Therefore it seems reasonable to hypothesise that if a person chooses to go home for lunch he is more likely to choose car over public transport than if he does not go home for lunch. On the other hand, the decision whether or not to go home for lunch must depend upon the travel mode used. To illustrate this, it is possible to repeat the same example given above, but in a different manner: given that a traveller is going to travel to work by public transport, he is less likely to go home for lunch than if he was travelling by car. This discussion makes it quite evident that the choice of mode and the choice between a single or double chain, between going home at lunchtime and not doing so, are each dependent on the other and therefore should be modelled using a simultaneous structure.

Whilst the choice of making a single or double chain is, for a given traveller, clearly dependent upon the chosen mode and the level of service offered by that mode, different workers will, as regards whether or not they go home at lunchtime, have different preferences independent of the level of service. These preferences will be a function of the socio-economic circumstances of the individual. For example, a worker is more likely to go home for lunch if he is married and his wife is at home than if she is not at home; he is more likely to go home if to do so does not entail changing his clothing and having a very thorough wash, or if he is not required to have lunch with business associates, and so on.

The modelling of the choice of whether to go home or not at lunchtime therefore requires the augmentation of the mode-choice utility function with variables expressing these characteristics. This task, however, was excluded from the present study, where only a limited investigation of double chains was undertaken.

Two types of model were estimated. The first has a structure identical to that of the single chain mode-choice model, with the exception that the model also explains the conditional mode-choice behaviour of the double chain travellers. That is, given that a choice has been made whether or not to go home at lunchtime, what are the mode-choice probabilities? The second type of model has a simultaneous structure which explains jointly the mode-choice and the lunch decision. The results of the conditional models estimated are given in Table 5.11. For a person who makes only a single chain – i.e. who does not go home at lunch time – the specifications of both the models given in Table 5.11 are identical to model 14 given in Table 5.2. For a person who makes a single chain the variable *LTT* in model 1 is assigned a value of zero; for one making a double chain this variable is assigned a value of the total time required for a work–home–work trip. Thus for car, bicycle and moped it is assigned the value equal to the sum of *IVTT* and *POVTT*, for walk it is assigned the value of *WOVTT*, and for public transport it is assigned the sum of the values of *IVTT*, *WSOVTT* and *SOVTT*.

Table 5.11

Estimation results: conditional double chains models

Variable	Model 1	Model 2
IVTT	–·0781	
	·0080	
WOVTT	–·0926	–·1512
	·0198	*·0234*
POVTT	–·0849	
	·0602	
WSOVTT	–·1254	–·1195
	·0223	*·0234*
SOVTT	–·0754	–·0807
	·0274	*·0284*
CAOD lnIVTT*	·0073	
	·0053	
FBOP	·5605	
	·1994	
BFMOA	–1·4655	
	·3787	
PTCON	1·5969	·5416
	·8178	*·8757*
LTT	–·0528	
	·0273	
*IVTT*1		–·0654
		·0091
*POVTT*1		–·4589
		·0862

Variable	Model 1	Model 2
*IVTT*2		–·0384
		·0130
*POVTT*2		–·1949
		·0704
CAOD lnIVTT*1		1·0504
		·1834
CAOD lnIVTT*2		–·0020
		·0012
*FBOP*1		1·2916
		·3101
*FBOP*2		·6957
		·5862
*BFMOA*1		·2982
		·5419
*BFMOA*2		–1·8152
		1·2436
$L^*(0)$	–558·71	–558·71
$L^*(\hat{\theta})$	–372·02	–343·07
χ^2 (d.f.)	373·37 (10)	431·26 (14)
ρ^2	.33	·39
$\bar{\rho}^2$	·33	·38
No. of observations: 511		No. of cases: 1,068

Model 2 given in Table 5.11 allows some of the variables to take different coefficients for single and double chains (as denoted by the suffixes 1 and 2). An interesting observation on the results of model 2 is that:

$$\theta_{IVTT2} = \tfrac{1}{2}\,\theta_{IVTT1}$$

and

$$\theta_{POVTT2} = \tfrac{1}{2}\,\theta_{POVTT1}$$

Since *IVTT*2 equals 2**IVTT*1, it seems that the effect of level of service could be represented by the same set of coefficients for both single and double chains. The fact that all the vehicle availability coefficients are smaller for double chains than for single chains is probably a result of the popularity of the walk mode for double chains. Since there were no observations of public transport being the chosen mode for double chains, the public transport constant *(PTCON)* was incorporated in the single chain utility function only.

Whilst many other specifications for this model could be postulated, no further work was done, partially because of the budget and time constraints and partially from the conviction that the double chain single chain choice must be considered simultaneously with the mode-choice.

For demonstration purposes a simplified simultaneous model was estimated and the results of this are given in Table 5.12. This model has the same specification as model 14 with the addition of a lunch constant *(LCON)*, which is assigned a value of one for a double chain and zero for a single chain. This representation is clearly not very satisfactory, since we

Table 5.12

Estimation results: joint double chains model

Variable	Model 1
IVTT	–·0851
	·0069
WOVTT	–·1372
	·0158
POVTT	–·2483
	·0367
WSOVTT	–·1385
	·0233
SOVTT	–·0735
	·0288
CAOD lnIVTT*	·0045
	·0056
FBOP	·2127
	·2218
BFMOA	·0712
	·4160
PTCON	·5166
	·8012
LCON	·1702
	·2386
$L^*(0)$	–854·72
$L^*(\hat{\theta})$	–559·85
χ^2 (d.f. = 10)	–589·73
ρ^2	·34
$\bar{\rho}^2$	·34

No. of observations: 511
No. of cases: 2,286

are endeavouring to represent in a single constant the preferences for a lunch trip. These are independent of the level of service variables but dependent on the socio-economic characteristics of the different travellers. The implication of *LCON* is that, given a chosen mode, the decision whether or not to go home for lunch has the same choice probabilities for all workers. Thus the model is badly specified, but it was intended primarily to demonstrate the differences between the conditional and the joint model and the desirability of the latter, given the *a priori* reasoning which justified a simultaneous structure. The models estimated with the double chains indicate the need to model this choice explicitly in those situations in which a significant proportion of double work chains are made.

5.10 Summary

The mode-choice model for work trips described in this chapter is probably the first attempt to consider the full variety of travel modes available in medium and small size Dutch communities, communities where the conventional binary-choice model (for car and public transport) commonly used in Britain and North America is clearly not suitable. While the models developed require further work before they could be considered to be standard operational production techniques, the existing models could serve usefully in various transportation planning studies. The study has also demonstrated that multinomial logit is a practical tool in the development of multi-modal models.

Whilst a number of conclusions can be drawn from the models estimated, it must be remembered that they relate specifically to the situation for which the models were estimated; the expectations that disaggregate models will prove to be more readily transferred from one area to another have not yet been proved. The first conclusion is that the probability of anyone choosing a given mode is largely determined by factors other than the level of service offered by that mode. If a car is perfectly available to a traveller, then there is a very high probability that he will choose it, regardless of the characteristics of the alternative modes. The policy implications of this, as proposals for traffic restraint in urban areas increase, at least in Europe, are considerable. The estimation results tend to confirm the general assumptions about the relative weights of in-vehicle and out-of-vehicle travel time, although it would appear that there could be significant differences in the evaluation of different types of out-of-vehicle travel time. In-vehicle travel time would, on the contrary, seem to be viewed similarly for all modes.

Travel costs do not apparently influence mode-choice, with the particular data-set, nor do socio-economic characteristics other than vehicle availability.

It has been previously suggested that the estimation of disaggregate models requires fewer observations than does the calibration of aggregate models. Estimating the same model with different data-sets tends to confirm this belief in that the marginal value of increasing the sample size above some 300 observations was found to be small.

The transformation of disaggregate models to aggregate models for use as predictive models presents a number of theoretical problems. It would seem possible, however, that, for all practical purposes, the effects of the problems can be minimised by the use of market segmentation, provided, of course, that the model specification is such as to make this possible.

Urban transportation studies are generally concerned with the prediction of total travel demand per facility over a period of at least ten years. Many policy issues are, however, related to the short term and, furthermore, do not require area-wide predictions. Much can be learned about travel demand, and the potential effect of various measures on that demand, through analysis of demand elasticities. A series of examples of direct, cross and arc elasticities have been developed to illustrate this.

The decision on whether to go home at lunchtime and the choice of mode for work trips can be expected to be highly interdependent. A conditional and a joint model have therefore also been developed to demonstrate the two different approaches.

6 The Shopping Destination- and Mode-Choice Model

6.1 Introduction

The shopping trip model developed during the course of the study is described and discussed in this chapter. The model developed was a joint destination-and mode-choice model. Thus, given the number of shopping trips originating from a residential area, the model can be used to distribute the trips over the alternative shopping destinations and also over the relevant alternative modes of transport.

It would seem desirable that a truly behavioural shopping model should have a simultaneous structure in which the choices of frequency, time of day, destination and mode are all considered. However, in view of the overall objectives and scale of this study, the scope of the shopping model developed had to be limited.

Traditionally, urban travel demand forecasting models treat the choices of frequency and time of day independently of the transport level of service, and thus independently of either destination or mode-choice. The data available from conventional urban transportation surveys are thus not generally suited to the development of frequency-choice models, nor are they suited to the satisfactory development of models in which the choice of the time of day at which a trip is made is explicitly modelled.

Given the resources available and the form of the models currently used in urban transportation planning studies, it was considered that a simultaneous destination-and mode-choice model provided the best scope for the introduction of immediate improvements to existing techniques. Furthermore, the implementation of a simultaneous model of destination- and mode-choice in software packages generally in use in urban transportation planning is likely to require fewer modifications than a more complex model in which frequency and time of day are also included.

A highly attractive feature of a simultaneous destination and mode-choice model is its similarity of specification to that of the Wilson model (1969), which is used extensively in urban transportation planning studies in the Netherlands and other European countries (see, for example, Buro Goudappel en Coffeng, 1972; MacKinder *et al.*, 1970); this similarity is discussed in section 7.2.

This discussion should not be taken to imply that frequency and time of day choices are not significant determinants of shopping travel patterns. On the contrary they are likely to be highly significant, but before they can be fully evaluated further data, and research and development are required.

Most of the work to date on the behavioural value of time in travel demand modelling has been concerned with home-based work trips, and more specifically with mode-choice for such trips. Travel demand forecasting for other purposes has therefore tended to be based upon the same values of time, or relative generalised price coefficients (see, for example, McIntosh and Quarmby, 1970).

The models described in this chapter together with the models described in Chapter 5 provide some very useful information on the validity of this approach.

6.2 The theoretical model

The model developed in this study is one which predicts the joint probability of choosing a destination and mode combination, given that a trip is made. Thus the dependent variable can be denoted as follows:

$$P(d,m:DM_t) \quad \text{or} \quad P_t(d,m)$$

where m denotes an alternative mode, d denotes an alternative destination and DM_t denotes the set of relevant destination and mode combinations available for a shopping trip to traveller t.

The logit model predicting this joint probability is written as follows:

$$P(d,m:DM_t) = \frac{e^{U_{dmt}}}{\sum\limits_{d',m' \epsilon DM_t} e^{U_{d'm't}}} \qquad (6.1)$$

where U_{dmt} is the utility to traveller t from going to destination d by mode m (see section 3.5.1).

The utility functions are assumed to be linear in the coefficients as follows:

$$U_{dmt} = X_{dmt}{'}\theta$$

$$= \sum_{k=1}^{K} X_{dmtk}.\theta_k \qquad (6.2)$$

where X_{dmt} is a vector of finite functions of the explanatory variables and θ is a vector of coefficients.

The alternative modes used in this study are: walk, bicycle, moped, car, bus, and train. The alternative destinations are those described in section 4.5.2.

6.3 Variables used in the models

The explanatory variables used in the shopping models include all those used for the work models and described in section 5.3. In addition, retail employment was used to describe the individual shopping centres; this was denoted as *EMP*. This variable can be considered as a proxy for the shopping opportunities available at the shopping centres. Other attraction variables which were available (see section 4.5.6), such as employment in those services which could make a shopping centre more attractive (banking, hairdressing, etc.), were not used in any of the models estimated.

6.4 Alternative specifications

The alternative specifications formulated for the shopping model were based to a great extent on the same sort of reasoning as was applied to the development of the work mode-choice model described in Chapter 5. The work on the development of the shopping model was commenced after the conclusion of that on the work mode-choice models and thus benefited from the results of that work.

The level of service variables were specified in the same manner as models 5 to 14 of the work mode-choice model – i.e. *IVTT, WOVTT, POVTT, WSOVTT,* and *SOVTT.* Thus, the alternative specifications tried differ only in the manner in which the vehicle availability and attraction variables were introduced.

Because of the distinct difference between the centre of Eindhoven and all the other local shopping centres – i.e. those in Son, Breugel and Best – some variables were introduced in some specifications twice; once specific to the Eindhoven centre and once specific to all the other, local centres. Variables specific to the Eindhoven centre are denoted by the prefix or suffix *EI* whilst those specific to the local centres are denoted by the prefix or suffix *LOK*.

In order to minimise potentially abortive work, any variable which had not been found to be satisfactory in the work mode-choice models was excluded from the specifications tried for the shopping model, with the exception of out-of-pocket travel costs.

6.5 Discussion of the estimation results

A full discussion of the results of the models estimated must take place within the context of the data set utilised. It is therefore necessary to consider the characteristics of the alternative shopping destinations, since those for the centre of Eindhoven are substantially different from those of all the other local shopping centres.

The data set could be described thus: consider a multi-dimensional space in which the axes are defined by the explanatory variables. For a given observation (i.e. a person making a shopping trip) an alternative destination and mode combination is described by a point in this space. An observation with a total of *DM* alternatives will be represented by *DM* points. For any observation in the data set utilised for this study, this procedure results in two clusters of points, and these two clusters are widely separated. One large cluster contains all the alternatives relating to the local shopping centres, and the second, smaller cluster contains the alternative modes for the Eindhoven centre. The peculiarity of the data set is that the Eindhoven cluster will be very much further away from the origin than the local cluster. This situation can be depicted in a simplified way as shown in Fig. 6.1. The distance between the two clusters is evidently very large if we consider, for example, retail employment and the average distance of the centre of Eindhoven from the residential areas. Not only is the centre of Eindhoven nearly 20 times as large as the largest of the local shopping centres in terms of retail employment (Tables 4.17 and 4.18) but the average distance from the residential areas to a shopping-centre is about ten times as great for the centre of Eindhoven as it is for the other, local centres. Furthermore Eindhoven is also the only centre to which trips were made by public transport.

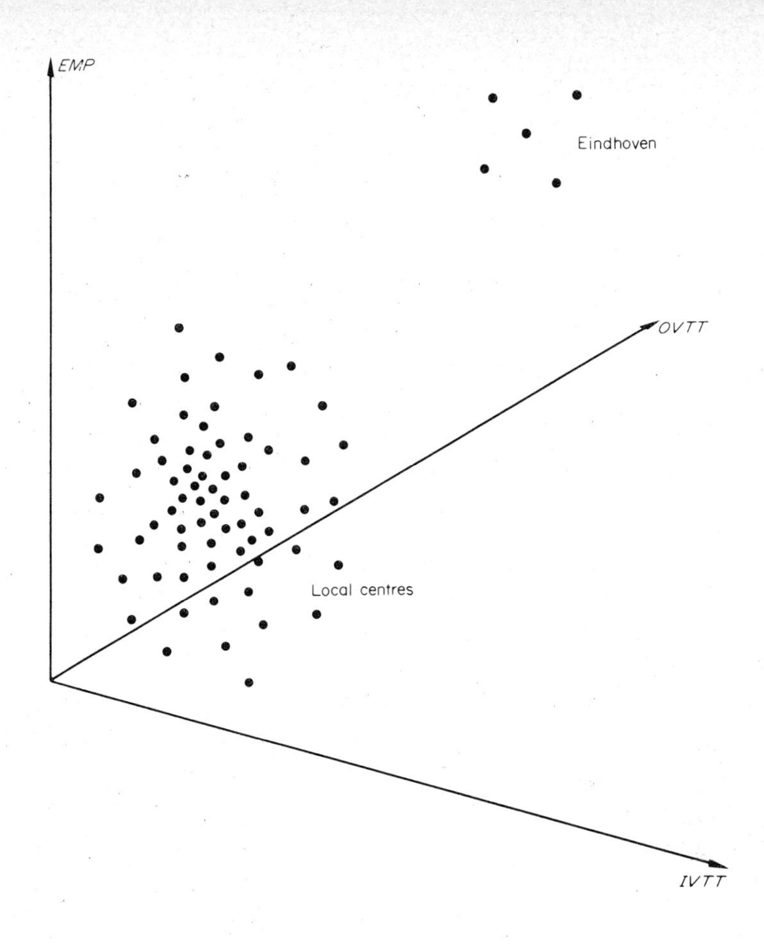

Fig. 6.1 A diagrammatic representation of the shopping data set

This structure of the data set suggests a hypothesis of a hierarchy of decisions: first, a choice between shopping in the centre of Eindhoven and shopping locally; and, secondly, if shopping locally is chosen, a choice among the local centres. The flaw in this hypothesis is that its reasonableness stems only from the fact that no medium size shopping centres existed in the sample. Indeed, a closer examination of the cluster of local centres will reveal a hierarchy of centres within this cluster. Thus any explicit modelling of a specific hierarchy will be arbitrary, while a model which considers all alternatives simultaneously will be more general.

Although this latter approach is sound in theory, the existing data set creates a specific problem in that the weight of the centre of Eindhoven can have a major influence on the determination of the model coefficients. In general, whilst it is desirable to have maximum variability in all of the variables considered, in this particular case a large portion of the variability is of a discontinuous nature, resulting from the fundamental differences between the local and Eindhoven alternatives. Unfortunately, it is impossible to determine the effect of this discontinuity on the estimation results without applying the model to a data set which exhibits more continuous characteristics. Although the problem of the relative sizes of the alternative shopping destinations cannot be resolved within the VODAR area, other problems of discontinuity could be resolved by taking a sample for a corridor or sector through the study area.

One consequence of the characteristics of the data set available is a lack of variability in the public transport services. Whilst the walk to the station variable *(WSOVTT)* has a reasonable degree of variability, the station waiting time variable *(SOVTT)* has extremely limited variability, because bus and train were used only for trips to Eindhoven; this is probably the reason why the coefficient of *SOVTT* in model 1 is not significant (Table 6.1).

Model 1 has an identical specification to that of the final work mode-choice model, model 14, with the addition of the retail employment variable *(EMP)*. As noted above, the coefficient of *SOVTT* was found to be insignificant, as was that of *PTCON*. A second model, model 2, was therefore estimated in which these two variables *(SOVTT* and *PTCON)*

Table 6.1

Estimation results of alternative joint models for shopping trips

Variable	Model 1	Model 2	Model 3	Model 4	Model 5
IVTT	—·1740	—·1766	—·1531	—·1502	—·1566
	·0116	*·0129*	*·0125*	*·0132*	*·0118*
WOVTT	—·1988	·1945	—·2122	—·2130	—·1951
	·0160	*·0152*	*·0177*	*·0184*	*·0157*
POVTT	—·4438	—·4268	—·4325	—·4694	—·3833
	·0486	*·0465*	*·0511*	*·0588*	*·0484*
WSOVTT	—·2282	—·2557	—·2605	—·2158	—·2350
	·0464	*·0232*	*·0246*	*·0237*	*·0275*
SOVTT	—·0148				
	·0689				

Table 6.1 continued

Variable	Model 1	Model 2	Model 3	Model 4	Model 5
OPTC					·0057
					·0136
*CAOD*lnIVTT*	·4885	·4816	·5734		·4035
	·1985	*·1989*	*·2197*		*·2129*
CAODEI				3·8236	
				·8803	
CAODLOK				1·1215	
				·5048	
FBOP	1·1429	1·1504	·7254	·8478	·7790
	·2289	*·2279*	*·2325*	*·2818*	*·2351*
BFMOA	−2·2901	−2·2119			
	·6092	*·6015*			
BFCON			−1·5163	−1·1389	−1·4550
			·2920	*·3139*	*·2906*
PTCON	·5789				
	1·9296				
EMP	·0024	·0024			
	·0001	*·0001*			
EMPEI			·0027	·0024	
			·0002	*·0002*	
EMPLOK			·0344	·0343	
			·0017	*·0018*	
lnEMP					·7419
					·0609
EICON					3·2645
					·8060
χ^2(d.f.)	695·7 (10)	694·5 (8)	1,290·6 (9)	1,282·4 (10)	794·4 (10)
ρ^2	·25	·25	·46	·45	·28
$\bar{\rho}^2$	·25	·25	·46	·45	·28

No. of observations: 432 No. of cases: 11,293

were omitted. From the results given in Table 6.1 it can be seen that this omission has negligible effects on the coefficients of the remaining variables; indeed the standard error of *WSOVTT* was reduced by 50 per cent.

The two variables *SOVTT* and *PTCON* typify the particular problems of the data set utilised: the lack of variability in *SOVTT*, and the fact it is an alternative specific variable applied only to those alternatives to which *PTCON* is also applicable, suggests that a high degree of collinearity could be expected. Furthermore, in so far as the observed choices are concerned, both these variables take a non-zero value only when Eindhoven is the chosen alternative.

The moped availability variable *(BFMOA)* has what, on *a priori* reasoning, seemed to be an illogical sign, since it suggests that as moped availability increases so the probability of choosing a moped decreases. *BFMOA* was therefore replaced by a simple moped constant *(BFCON)* in models 3, 4 and 5. The problems encountered with the variable *BFMOA* in both this model and the work mode-choice model clearly indicate that a much more detailed examination of the characteristics of the moped user is required. This is especially important in view of the popularity of the moped in the Netherlands.

In view of the discontinuity problems described above, the retail employment variable *(EMP)* was split into two variables in model 3. The one was specific to the local centres *(EMPLOK)* and the other to the centre of Eindhoven *(EMPEI)*; the Eindhoven specific variable was thus essentially a constant but taking the value of the retail employment in Eindhoven. This change gave a significant improvement in the goodness of fit in comparison with models 1 and 2, with $\bar{\rho}^2$ increasing from 0·25 in model 2 to 0·46 in model 3. This change indicates that the difference between the local centres and the centre of Eindhoven was not captured by a linear *EMP* variable.

The effects of the replacement of a single employment variable by two illustrate some of the particular problems which can be associated with a data set displaying characteristics such as the one used in this study.

The local employment variable, *EMPLOK*, has a very much larger coefficient (0·0344) than the Eindhoven employment variable, *EMPEI*, (0·0027), although the coefficient of the joint employment variable, *EMP*, in model 2 is 0·0024. The fact that the coefficient of the joint variable *EMP* has a similar value to that of the Eindhoven employment variable, *EMPEI*, indicates the weight of the Eindhoven alternatives in the determination of the coefficient of the joint variable.

The difference between the coefficients of *EMPLOK* and *EMPEI* could be interpreted as representing an effect of diminishing marginal returns of shopping centre size. The attractiveness of a centre should thus be specified in the utility function so that it increases less than linearly with increasing employment. This means that if a single employment variable is used in the utility function it should be a concave transformation, such as *lnEMP* as used in model 5. The statement about diminishing returns of size applies only to the utility functions. Therefore the diminishing effect of size on the utility does not necessarily imply a diminishing effect in terms of actual trips per employee. This latter effect is also a function of the actual choice probability and can be determined only by using the actual values of the variables.

If a variable is introduced in the additive utility function in its natural log transformation, then the elasticities with respect to this variable are not a function of the value of the variable itself, but of its coefficient and the choice probabilities (see section 3.5.4). Therefore the coefficient of 0·74 of *lnEMP* in model 5 implies that the choice probability is always inelastic with respect to employment, since a 1 per cent increase in size results in less than a 1 per cent increase in the probability of that alternative being selected. Thus, the larger the centre, and hence the greater the probability of choice, the more inelastic its attractiveness with respect to size.

The basic specification of model 3 was maintained for model 4, in which the form of the coefficient of *CAOD* utilised in model 14 of the work mode-choice model was examined in further detail. This was done by replacing the one variable *CAOD lnIVTT* by two separate variables, *CAODEI* and *CAODLOK*, the one being specific to Eindhoven and the other to all the local centres. In terms of both goodness of fit and the individual coefficients there does not seem to be a major difference between models 3 and 4, but the values of the two *CAOD* coefficients in model 4 demonstrate very clearly the variability of the *CAOD* coefficient with trip length. This confirms the reasoning applied in the development of the work mode-choice model and implemented through the multiplication of *CAOD* by *lnIVTT* (see section 5.6). The ratio of the coefficient of *CAODEI* to that of *CAODLOK* is of the same order as the ratio of *lnIVTT* for Eindhoven to that of the average value of *lnIVTT* for the local centres.

In models 3 and 4 the inclusion of variables specific to the centre of Eindhoven made it inadvisable to include a public-transport constant; furthermore this was found to be insignificant in model 1.

In model 5 the car availability variable, *CAOD*, was returned to the form in which it had been included in models 1 to 3 – i.e. *CAOD.lnIVTT* and the *EMP* variable was introduced in its natural log transformation, *lnEMP*. Although out-of-pocket travel costs, *OPTC*, had been found to be insignificant in the work mode-choice models, it was decided to introduce them in model 5; a similar result was, however, obtained.

An Eindhoven constant, *EICON*, was introduced to account for any pure alternative effect, i.e. any model specification deficiencies. This constant was found to be highly significant, thus suggesting that *lnEMP* does not explain all the differences in attractiveness between the centre of Eindhoven and the local centres.

The positive coefficient of *EICON* indicates that Eindhoven centre has an added attractiveness not captured by the retail employment variable.

Whilst model 5 would seem on *a priori* reasoning to have the best specification of those tried, the goodness of fit is very much lower than for either model 3 or model 4. The coefficients of all the level of service and vehicle availability variables are, however, the same. The only differences between the two models is the manner in which the employment variable was introduced, and the inclusion in model 5 of an Eindhoven constant (because of perfect collinearity with *EMPEI* such a constant could not have been included in model 3) and out-of-pocket travel costs.

6.6 Comparisons with the work mode-choice

The shopping model described in this chapter is a simultaneous model of mode- and destination-choice, whilst the work model described in Chapter 5 is a conditional model of mode-choice given the destination. In order to provide a more direct comparison between the two models, a conditional mode-choice model given the destination was also estimated for the shopping sample. This model was estimated with a specification identical to that of model 14 for the work mode-choice model. The estimation results for these two models are given in Table 6.2

From this table it can be observed that the coefficients of *IVTT* are similar for both models, while all the coefficients of the out-of-vehicle travel time variables are greater for shopping than for work. This would appear to be reasonable, since it could be expected that out-of-vehicle travel time, relative to in-vehicle travel time, will be regarded as more inconvenient in the case of shopping trips than in that of work trips, not only because of the need to carry the purchases but also in relation to the total duration of the home-to-home journey. The greatest increases in the weights of the out-of-vehicle travel time elements are in the

Table 6.2

Estimation results of conditional work mode-choice model 14 and conditional mode-choice model for shopping

Variable	Work	Shop
IVTT	−·0600	−·0551
	·0093	*·0151*
WOVTT	−·1192	−·1413
	·0295	*·0221*
POVTT	−·3260	−·3850
	·0946	*·0714*
WSOVTT	−·1136	−·2154
	·0234	*·0684*
SOVTT	−·0856	−·3144
	·0288	*·2372*
*CAOD*lnIVTT*	1·0056	1·7671
	·1800	*·3901*
FBOP	1·4348	0·8167
	·3211	*·2796*
BFMOA	·6689	−1·1521
	·5536	*·6456*
PTCON	·5057	4·8982
	·9252	*3·4747*
$L^*(0)$	−435·13	−412·99
$L^*(\hat{\theta})$	−260·54	−311·35
χ^2 (d.f. = 9)	349·17	203·28
ρ^2	·40	·25
$\bar{\rho}^2$	·33	·24
No. of observations:	390	432
No. of cases:	845	724

coefficients of the two public transport specific variables, *WSOVTT* and *SOVTT*. This comparison can have important implications in travel demand modelling, since it indicates that there could be no justification for the use of a unique set of generalised price coefficients across all trip purposes, a conclusion also reached by Watson (1974).

A further important comparison which should be made is between the conditional mode-choice shopping model and the joint destination-and mode-choice shopping models (model 3). This is done for the level of service variables in Table 6.3. The significant differences between the two models are in the coefficients of *IVTT* and *WOVTT*. The increase in the absolute values of the coefficients in the joint model relative to the conditional model could be attributed to the effect of distance on shopping destination-choice.

Table 6.3

Comparison between the estimated coefficients of level of service variables for a conditional mode-choice model and a joint destination- and mode-choice model (model 3) for shopping.

	Shop $P(m\|d)$	Shop $P(m,d)$ model 3
IVTT	−·055	−·153
	·015	*·013*
WOVTT	−·141	−·213
	·022	*·018*
POVTT	−·385	−·433
	·071	*·051*
WSOVTT	−·215	−·261
	·068	*·025*

This effect could be a result of a tendency to visit nearby shopping centres which are more familiar. Thus these differences could be a result of not including a familiarity variable in the joint model (distance could have been a proxy). However, no model is perfectly specified and it is also possible to assume that the coefficient of *IVTT* in the conditional model is biased due to mis-specification of modal captivity. Although it is not clear which choice is mis-specified more severely – destination-choice of mode-choice – it is clear that, given the differences between the joint and the conditional model, and given the *a priori* assumption of a simultaneous structure, the results of the joint model are more reliable and should in general be used.

6.7 Summary

A high degree of interdependence between the choice of destination and mode for shopping trips can usually be expected. These decisions should therefore ideally be modelled in a single simultaneous model, such as those described in this chapter. Although the data set used has a number of particular characteristics, the estimation results suggest that the basic modelling philosophy is of considerable potential. The attractiveness of the alternative shopping centres was found to be better described using the natural log of retail employment rather than retail employment as a linear variable. The effect of car ownership on mode-choice was found to be a function of travel time, as was also the case with the work mode-choice model.

A comparison between a mode-choice model for shopping and one, using the same specification, for work suggests that the relative weights of in-vehicle travel time and the various elements of out-of-vehicle travel time cannot be assumed to be constant across all trip purposes.

7 A Review of the Study and the Models

7.1 Introduction

The purpose of the study was to examine the viability in a Dutch context of disaggregate and simultaneous travel demand modelling. Whilst the theoretical advantages of both disaggregate and simultaneous demand modelling have been well expounded over recent years (see, for example, HRB, 1973; Ben-Akiva, 1973; CRA, 1972; de Donnea, 1971; Fleet and Robertson, 1968: Oi and Shuldiner, 1962) and are explained in Chapters 2 and 3 of this report, it is relevant to recapitulate briefly.

The aggregation of data over a group of individual people, be the form of aggregation geographic or on the basis of some other characteristic, tends to conceal differences between individuals which can be of great significance. This is because the groups of people formed by aggregation are generally represented, either explicitly or implicitly, both in model estimation and forecasting by the average value of each variable. Thus neither the range of the values of the variables nor the distribution of those values are recognised, except in a very simplistic way.

The averaging effect of aggregation can have two major effects on transportation models. One is that the models based on aggregate data are unlikely to describe the true demand for travel since they are not directly concerned with the behaviour of the individual consumer. Furthermore, by aggregating, the full range of possible combinations of relevant characteristics is unlikely to be present in the aggregated set of data on which the model was based. Although an aggregate model might be able adequately to simulate the observed aggregate situation from which it was derived, its stability over different forms of aggregation, and thus its time and geographic stability under the practical conditions of transportation studies, is very definitely open to question. An aggregate model is much less likely to be a truly behavioural model than a disaggregate model. The second of the effects of aggregation is the need for more sets of individual observations than would be necessary with a disaggregate model because of the effective loss of information implied by aggregation. Thus disaggregate models should not only be technically better than aggregate models, but they should also be more responsive to changes in society and transportation systems, have greater time and geographic stability, and require fewer sets of data.

The conventional urban transportation model structure is recursive – that is, it consists of a series of sub-models applied sequentially (thus the output of one model is utilised in the input to the following model) without feedback between sub-models (de Neufville and Stafford, 1971). This implies a conditional, or step-wise, decision-making process. It is thus assumed that the decision to make a trip for a particular purpose is independent of either the alternative destination or mode chosen, and that the choice of destination is independent of either the frequency with which the trip is made or the chosen mode. For many trips this conditionality – that is, given a destination, which of modes 1 . . . M is chosen – can clearly be faulted. The decision on whether to make a trip or not, on where to go and how to go, is in fact usually not a series of separate decisions but one complex, or joint, decision in which destination d and mode m are simultaneously selected from the complete set of alternative destinations and modes, DM.

Simultaneous models which endeavour to capture, or to explain, this complex decision in a single, joint model, can be expected to be closer to reality than recursive models which artificially attempt to isolate into a series of consecutive, conditional decisions what is really a single complex decision.

Disaggregate models can be either recursive or simultaneous, and simultaneous models can be either disaggregate or aggregate. As a general rule, however, the ideal system would seem to be that of disaggregate simultaneous modelling.

In the study, both a simple disaggregate conditional model and a disaggregate simultaneous model were developed utilising data from the Eindhoven area. The function of this chapter is to examine this work in relation to conventional urban transportation models, in relation to issues currently confronting those concerned with transportation planning and policies, and in relation to both the data available and ideal data requirements. Based on these analyses, possible implications for the future development of procedures for the analysis and forecasting of passenger travel demand are identified.

7.2 Model performance in comparison with conventional models

The results of the work described in Chapters 5 and 6 clearly demonstrate that the modelling procedure adopted gave satisfactory results.

The specification of the home–work–home mode-choice model utilised in this study is basically of similar form to the modal split derived by Wilson (1969), which is one element of what is possibly one of the most advanced trip distribution/modal split models in common usage in production studies. We have therefore selected it as the model with which to compare the model structure and modelling strategy adopted in the study. Wilson's modal split model can be specified thus:

$$\frac{T_{ijm}}{\sum_{m' \epsilon M} T_{ijm'}} = \frac{f(C_{ijm})}{\sum_{m' \epsilon M} f(C_{ijm'})} \qquad (7.1)$$

where T_{ijm} = the number of trips from i to j by mode m;
C_{ijm} = the generalised cost of travelling from i to j by mode m.

If we assume that $f(C_{ijm}) = e^{\beta C_{ijm}}$, as is a common convention, then we have:

$$\frac{T_{ijm}}{\sum_{m' \epsilon M} T_{ijm'}} = \frac{e^{\beta C_{ijm}}}{\sum_{m' \epsilon M} e^{\beta C_{ijm'}}} \qquad (7.2)$$

which is a share model similar to the multinomial logit model:

$$P(m:M_{dt}) = \frac{e^{U_{mt}}}{\sum_{m' \epsilon M_{dt}} e^{U_{m't}}} \qquad (7.3)$$

The primary difference between logit and Wilson's model lies in their respective theories. Wilson's model is based on the concept of entropy, whilst the logit model is based on individual utility theory. Wilson's model is thus an aggregate model and logit is a disaggregate model. The second major difference between these two models lies in the estimation of the generalised cost function, or the utility function. In the conventional use of the Wilson model the coefficients of the elements composing the generalised cost are determined externally (see, for example, McIntosh and Quarmby, 1970), and in the estimation, or calibration, of the modal split model given in (7.2) only the coefficient β is determined (e.g. Wilson, Hawkins *et al.*, 1969). In the multinomial logit model all the coefficients of the variables composing the utility functions are usually estimated during the calibration process. The estimation, or calibration, procedure therefore provides a high degree of flexibility in the choice (and evaluation) of variables for incorporation in the utility functions. Thus, for example, many socio-economic variables can be readily included, whereas they can, conventionally, only be included in a Wilson type model by use of some stratification scheme by person type, allowing the range of alternatives to vary by person type as well as the coefficients of the generalised cost function.

It is pertinent to note that the Wilson model, and thus the logit model, can both be compared with a simple diversion curve procedure since, utilising Wilson notation:

$$\frac{T_{ij1}}{T_{ij2}} = \frac{e^{\beta C_{ij1}}}{e^{\beta C_{ij2}}}$$

$$= e^{\beta(C_{ij1} - C_{ij2})} \qquad (7.4)$$

where suffixes 1 and 2 indicate modes 1 and 2. Thus, the ratio of the volume of trips by one mode to that by another mode is equal to the ratio of functions of the travel costs by each mode, or a function of the difference in travel costs.

The multinomial logit model applied to a conditional modal-choice problem can thus be compared in structure to some extent with modal split models in current use. It does, however, offer major advantages over both of these models. The first of these is that the model can be used with disaggregate data, with individual observations, thus avoiding all the problems of averaging implied by aggregation. A consequential advantage of this is the low number of observations required for model estimation. In this particular study, in which only 390 observations of valid home–work–home single chains were available, it proved possible to estimate a model which satisfied all the major criteria of model building. All coefficients had a logical sign, the relative values of comparable coefficients were not such as to suggest that the individual coefficients were likely to be significantly in error, and, furthermore, the standard errors of the level of service variables, as well as the socio-economic variables (moped ownership, *BFMOA*, excluded) were small. It should be recorded, however, that although 390 observations appeared to be an adequate number in this specific instance, a similar number need not be adequate under any other circumstances.

A second, consequential advantage of the combined use of disaggregate data and the logit model is the ability to recognise within the model structure that different people have different sets of alternatives available. Not only does this have major implications for computational economics, but it also means that a model thus estimated has a much greater chance of being truly behavioural than a model structure in which this characteristic is not present.

The fairly extensive experience of one of the authors with Wilson type models (as aggregate models) suggests that the number of observations utilised in this study would have been inadequate for satisfactory model calibration with even two modes, let alone the six for which the models in this study were estimated. Furthermore, as already stated, calibration involves estimation of only one parameter, the relative values of the coefficients of the various level of service variables having been determined externally, or taken from a manual. Yet model 14 of the home–work–home mode-choice models described in Chapter 5, when applied as an aggregate forecasting model, was found to reproduce the observed situation quite adequately given the relevant circumstances (section 5.7.2). With the exception of the vehicle availability variables *(CAOD, BFMOA* and *FBOP),* the variables in most of the home–work–home mode-choice models estimated in this study are comparable to those utilised in many applications of the Wilson model, both in Britain and by Buro Goudappel en Coffeng in the Netherlands. In conventional British studies, for instance, it is standard practice to assume that:

Generalised Cost = A (access + waiting + transfer time)
$+B$ in-vehicle travel time
$+C$ out-of-pocket travel cost

and, furthermore, that $A = 2B$, that both are a linear function of the value of time, and the $C = 1{\cdot}0$ – in practice, coefficients are frequently divided by the value of time to give units scaled in time (McIntosh and Quarmby, 1970). This implies a predetermination of the trade-offs of in-vehicle travel time against out-of-vehicle time as well as of the value of time. The public transport constant utilised in the models developed in this study can be compared with the public transport handicap proposed by Wilson, Hawkins *et al.* (1969), the use of which has become normal practice in British urban studies (for instance, MacKinder *et al.,* 1970). The stratification by person type in the Wilson model can be compared with the definition of available alternatives in disaggregate models of the logit type. It can also be compared with market segmentation in applications of logit models as aggregate prediction models. In the particular case of model 14, the car availability variable, *CAOD*, not only differentiates between different levels of car availability but also facilitates the use of the model as an aggregate forecasting model with the market segmented into persons with a car available and those without. If the market is to be segmented, then the model specification must be such that the choice probabilities vary between the different groups of travellers represented by the market segments. The inclusion of other vehicle availability variables such as *BFMOA* and *FBOP* permits yet finer forms of market segmentation.

The need to stratify the model into at least two market segments (car owners and non-car owners) to obtain acceptable aggregate predictions (section 5.7.2) illustrates the potential dangers of aggregation. Aggregation of the total population of a zone into one class, and assignment of the average zonal level of car ownership, implies that a car is being made available to everyone, regardless of whether they indeed had a car or not. Stratification of the model into market segments, or person types as proposed by Wilson, readily overcomes this problem. This problem, that of averaging across two or more unlike classes, can be of equal relevance to many other elements of transportation modelling.

Thus it seems reasonable to conclude that the combination of disaggregate data and the multinomial logit model makes it possible to estimate a mode-choice model for more modes with considerably fewer observations than would be required for a conventional aggregate Wilson type model, and simultaneously to determine local values of the coefficients of the model, such as those used in the Generalised Cost formula. Had the project been concerned with simply estimating a model composed of the same variables as used in a conventional application of the Wilson model, then it would not have been necessary to estimate more than that one model; this would have been a relatively quick and economic procedure. But the modelling strategy adopted for this study, and thus applicable to production studies too, permits the evaluation and inclusion of a whole range of variables not conventionally included in applications of the Wilson model, but which are relevant to mode-choice.

Disaggregate mode-choice modelling using the multinomial logit model can thus be seen as a major improvement on the conventional procedures utilised with the Wilson model, whilst being of similar form. Since the Wilson model itself is a major advance on the procedures used in many studies, the approach to modelling described in this report can form the basis for significant improvements in mode-choice modelling.

The results of the extension of the home–work mode-choice model to a joint number of chains and mode-choice model indicate the problems which can occur when a complex, or simultaneous, decision procedure is factored into two separate, or conditional, processes. If there are good reasons for believing that a decision structure is simultaneous, then the model structure should reflect this. To factor a simultaneous decision into two or more independent decisions can clearly lead to different results and, possibly, errors. Equally, if there is evidence to suggest that a decision structure is recursive, then it is wrong to model it as a simultaneous decision. It can also be wrong to estimate a joint (i.e. simultaneous) model and to apply it as a conditional model.

We believe that a high degree of simultaneity can be expected between choice of mode for work trips and the decision whether or not to go home at lunchtime. It is thus preferable to model these two decisions as a simultaneous decision. The fact that the coefficients of the variables in the comparable joint and conditional models are not similar and, therefore, that the joint model applied as a conditional model will not always give the same probabilities as the conditional model, could be attributed to faulty model specification. It could also be attributed, in part, to chance differences resulting from sample sizes and sampling procedures.

These differences strengthen the argument for simultaneous rather than recursive models when there are good *a priori* reasons for believing a degree of simultaneity to exist in the decision process. Thus, for most trip purposes the choice of destination and mode could be expected to be highly simultaneous, and the second series of models estimated are simultaneous destination- and mode-choice models for home–shop–home chains.

The structure of the destination- and mode-choice model utilised for this study can be compared with an origin constrained form of the Wilson type distribution model, which has the form:

$$T_{ijmn} = \frac{O_{in} D_j f_n(C_{ijm})}{\sum\limits_{j \epsilon J} \sum\limits_{m \epsilon M} D_j f_n(C_{ijm})} \qquad (7.5)$$

where T_{ijmn} = the number of trips from i to j by mode m for person type n;

O_{in} = the number of trip origins at i for person type n;

D_j = the number of trip destinations at j;

$f_n(C_{ijm})$ = a person type n function of the Generalised Cost of travel from i to j by mode m.

This can also be written:

$$\frac{T_{ijmn}}{O_{in}} = \frac{e^{lnD_j + ln\left[f_n(C_{ijm})\right]}}{\sum\limits_{j\epsilon J}\ \sum\limits_{m\epsilon M} e^{lnD_j + ln\left[f_n(C_{ijm})\right]}} \tag{7.6}$$

or if $f_n(C_{ijm}) = e^{\beta_n C_{ijm}}$, as is frequently assumed, then:

$$\frac{T_{ijmn}}{O_{in}} = \frac{e^{lnD_j + \beta_n C_{ijm}}}{\sum\limits_{j\epsilon J}\ \sum\limits_{m\epsilon M} e^{lnD_j + \beta_n C_{ijm}}} \tag{7.7}$$

which is a logit type specification, very similar to that of model 5 of the shopping models described in Chapter 6. In (7.7) the coefficient of the destination variable is equal to unity; this is necessary in the case of a model estimated using geographic aggregations of alternative destinations and if the model is to be applied to zoning schemes other than that for which it was calibrated (see section 2.2).

The points raised in the discussion on the simple mode-choice model are equally applicable to the joint destination- and mode-choice model. Using disaggregate data and the multinomial logit model it is possible to estimate a destination- and mode-choice model essentially comparable with that of the origin constrained Wilson type distribution/modal split model, but with many fewer observations. The lack of aggregate prediction tests in this study means, however, that this conclusion cannot be so definite as it was with the home–work mode-choice model. As with the mode-choice model, the ability to include socio-economic variables in the utility function adds considerably to the potential power of disaggregate modelling utilising multinomial logit.

Thus the disaggregate conditional and disaggregate simultaneous models estimated using the multinomial logit model offer major improvements over most conventional urban transportation modelling procedures, even those based on the Wilson model. The advantages are two-fold. First, with the same range of variables, the various coefficients can be estimated more readily, with fewer observations and probably with a greater degree of reliability; thus they offer a potential saving in either time or cost, or both. Secondly, they offer the opportunity to include more variables, and thus, for a similar cost, to develop models with more complete specifications, and, therefore, better models.

For prediction purposes, models estimated as disaggregate models have to be used as aggregate models, but as such they should perform as well as, and probably considerably better than, models estimated, or calibrated, with aggregate data. Among the major benefits offered by models of the type described in this report are those of utilising abstract mode specifications, of market segmentation through the inclusion of socio-economic variables and the ability to utilise distributions of variables rather than average values. A further, and very considerable advantage, is the independence of the model from zoning schemes, and thus the ability to apply one model to many different levels of geographic aggregation, according to the requirements of the specific problem being studied.

7.3 Relevance of the models to current issues

Among the most important practical criteria against which models can be judged are their relevance to current issues and the ease and speed with which they can be applied to evaluate the consequences of different policies or changing conditions (see Chapter 1).

The relevance of a model specification is, of course, dependent upon the ability to quantify a particular issue and to identify that quantity in the available data. The models developed in the course of this study all include both in-vehicle and out-of-vehicle travel time as variables, with the latter composed of a number of separate elements. Within the limits of the situation for which the models were estimated they are thus suitable for the evaluation of those effects of changes in the level of service which can be quantified in terms of travel time.

Out-of-pocket travel costs were, however, found to have no significant influence on mode choice for the data used. The models therefore cannot be utilised in their present form to evaluate the effects of changes in transport costs. Since transport costs are very much a current policy issue, it could be argued that the models are seriously deficient in this respect. But if they are deficient it is unlikely to be a result of the modelling procedure, since what the model indicates is that the cost of transport plays an insignificant role in the choice of mode in relation to the other variables included in the models, for the data from which the models were estimated. It could be argued that the *a priori* reasoning that cost should be included in a model specification – as, for instance, in the conventional Generalised Cost formulation described in section 7.2 – is wrong, and that a strength of the form of the modelling procedure utilised in this study is that it can refute such reasoning.

Nevertheless, one of the questions which has to resolved is whether this particular aspect of the model is attributable in some way to the data, or rather to a fundamental error in the *a priori* reasoning which leads to the inclusion of cost in many travel demand models, without any statistical tests but based on predetermined 'values of time'. Furthermore, even if costs are not very relevant now, they can be expected to become more relevant in the future as energy prices increase. However, work that Cambridge Systematics have carried out under contract to the US Department of Transportation, since completion of this study, utilising the same model estimation procedure, indicates that it is possible to introduce a cost variable using *a priori* information which increases the relevance of the model to current policy issues (Ben-Akiva and Lerman, 1974).

Questions such as the effect of comfort on the number of people using public transport cannot be answered by any of the models estimated, but this is primarily a question of availability of data. If data were available for an area in which a variety of levels of comfort were offered by public transport, then, even if those different levels could not be described quantitatively, the relative effect of them could be explained through the use of alternative specific qualitative variables (i.e. binary or dummy variables), provided that the level of comfort affects travel demand.

The very real current issue of restraint on private car usage through parking controls, although frequently described in terms of parking capacity, is in fact experienced by the individual traveller through parking charges, time taken to find a parking space, or travel time from a parking space to the ultimate destination, or any combination of these variables. Potentially the models developed are thus relevant to the issue of restraint, since all the time-based variables can be specified.

Another important issue in some areas is that of the effects of the introduction of a new mode of transport. Indeed this is particularly relevant in the Eindhoven area, where the potential suitability of the Krauss-Maffai Transurban personal rapid system is currently being studied. So long as the variables included in the model are both relevant and generic (i.e. so long as it is an abstract-mode model) then the model form used in this study is highly suitable for the evaluation of new modes. Thus, in this particular case, the effects on mode-choice of the potential improvements in level of service offered by Transurban could be determined. However, the effect of increased privacy or comfort over existing train or bus services could not be properly estimated, nor could it with other models in current use.

The ease with which the likely effects of different policies, or conditions, could be estimated is relatively great; many of these can be directly determined by the use of point or arc elasticities as described in Chapter 5.

The ease with which estimates of total movements are made is primarily dependent upon the facility with which the relevant data can be prepared, as it is with any other modelling procedure. The actual calculation of demand, given the necessary input data, is relatively simple and straightforward.

The use of disaggregate data in model estimation should give the models a greater degree of geographic stability than conventional aggregate models, and thus decrease the dangers of applying a model – or coefficients from a model – estimated for one locality to another locality. The final, but perhaps most telling, remark on the relevance of this modelling procedure to current issues is that it is, so far as we are aware, the only behavioural multi-modal mode-choice model estimated to date in which all the transport modes in use in medium sized Dutch towns have been included as independent modes.

Thus it can be concluded that this modelling procedure can be used to develop models sensitive to current issues; furthermore we can conclude that the models estimated in this study are probably considerably more relevant to current issues than the traditional models in use in transportation studies.

7.4 The data available

The work undertaken in this study had three main constraints: time, budget and data. The data constraint manifests itself in the volume and the quality of the data. The constraint on overall volume of data was a real constraint only for budgetary and time reasons. What was a problem, however, was the small number of observations in the sample initially selected in which public transport was the chosen mode (32 for home–work–home single chains and 21 for the shopping model) and, in the case of the shopping model, in which moped was the chosen mode (only 21 trips).

An additional problem in the shopping model was the geographic distribution of shopping centres apparently perceived by residents of Best, Son and Breugel as real alternatives. The fact that Eindhoven was the only alternative centre to those in the immediate locality requires cautious interpretation of the model. Eindhoven was considerably larger than any of the alternative centres, having some 3,700 employees compared with a range of 1 to 206 in the remaining centres; it was also the only centre for which public transport was reported as the chosen mode, and it was the only centre more than some 3km from a home. This particular problem had its compensations, though, since it provided an opportunity to obtain a somewhat greater insight as to the most likely form of some of the explanatory variables through the use of different model specifications.

Since there is only limited variability in the frequency of the public transport services between Best, Son and Breugel, and Eindhoven, the real effect of frequency of service on mode-choice might not have been determined in this study. A greater variety of frequencies is clearly desirable; variety in, or a wide range of, any variable is highly desirable for the estimation of a reliable, or robust, model.

The fact that the models were estimated for two particular residential areas with a very specific, bi-modal distribution of trip lengths could restrict the general applicability of the models.

The available information from the survey was originally not expected to be a problem, but during the course of the study a number of deficiencies was determined. One of the most important of these was a simple definition of moped availability: the variable moped ownership per member of the household aged over 15 years *(BFMOA)* was generally felt to be inadequate, but a real alternative could not be readily identified. Possibly a fine division of age over 20 years could have been of some value. Further information on both car parking (distance walked between address visited and car park, parking charge, etc.) and the type and ownership of the car used or available (company car or not) could have improved the estimation of the level of service variables and out-of-pocket costs.

This problem also relates to that of the quality of the level of service data, all of which had to be estimated, since no suitable data were directly available for the study area. Whilst the problem was no greater than it would have been with a conventional model, the deficiencies experienced are of much greater significance in that they could prevent the realisation of the full potential value of disaggregate models.

A major problem in the interpretation and evaluation of the model results is that we were not aware of any similar study with which the work undertaken in this study could be compared. Studies such as those by Ben-Akiva (1973) and CRA (1972) refer to American data and have few points in common with this study, whilst those based on European data, such as Watson's work (for example, 1974) and de Donnea's (1971), deal with very different situations.

7.5 Ideal data requirements

Whilst the use of disaggregate models, and especially disaggregate simultaneous models, with the data available from conventional urban transportation studies will permit the estimation of better models than those normally in current use, even greater improvements could be realised if additional data were available. By 'additional data' we mean more information and not more observations. Both the theory of the techniques and the practical experience gained in the course of this and other studies show that considerably fewer observations are necessary than with conventional aggregate models.

7.5.1 *Level of service data*

A much greater emphasis is required in future studies on the collection of level of service data than has been the general practice to date in Dutch transportation studies. Two main classes of data can be identified: that obtained during the course of a home-interview survey, and that requiring special fieldwork. Additional data that should be collected in the home interview survey are:

- type of ticket (e.g. single, day-return, six-trip, monthly season, etc.);
- location of parking place;
- whether the parking place available is a reserved space or not;
- parking charge;
- type and age of car available;
- number of passengers in the car;
- what car usage costs, if any, are shared with the passengers;
- what car usage costs, if any, are paid by the employer.

One of the problems associated with the use of level of service variables in travel demand models is the assessment of the values of those variables. It can be argued that travel decisions are not based on actual, observed, times and costs but rather on the expected, or perceived, values. Thus it could be argued that a truly behavioural model should utilise perceived values. In practice this presents major problems, which – though implicit in all forms of models – tend to be emphasised by the use of disaggregate models, which demand the explicit consideration of individual data. These problems are those of assessing for each individual traveller not only an unbiased estimate of the perceived times and costs associated with the chosen alternative, but also of the times and costs of the alternatives available but not chosen (see, for example, O'Farrell and Markham, 1974). A particular problem here is that of post-purchase bias in which a consumer, when questioned on the attributes of his available alternatives, tends to justify his choice by making it seem more attractive than the reality in comparison with the alternatives rejected. It is also quite possible that the perceived values of the LOS variables for a given alternative vary with the chosen alternative. Thus, for example, the perceived values for bus could differ between those who chose car and those who chose bicycle, both having rejected bus. This could have major implications for model structure.

A further, and highly relevant, problem is that of predictions. If we assume that the objective of most travel demand models is to prepare predictions of future travel demand, then we clearly need to be able to estimate the future values of the variables included in our models in a manner consistent with the base-year values. In the case of level of service variables we can only make objective, or engineering, estimates of future values. In order to be able to use perceived values in the model, we must have some mapping procedure whereby we can convert the future objective values to perceived values. An alternative procedure is to utilise only objective values throughout, both in model estimation and model application. The mapping between perceived and actual values thus becomes implicitly incorporated in the model. If the model is adequately specified then the difference between the use of perceived and objective values could be unimportant.

Thus, whether perceived or objective values are used in the model, objective values are needed for the base-year so that estimates can be made of perceived values for predictions from the objective data available.

Special surveys designed to collect level of service data, which should be conducted as an integral part of transportation studies, are primarily of the journey speed type, but special emphasis should be given to detailed studies of waiting and transfer times for public transport. Since waiting times are usually a function of frequency and reliability, but do not necessarily explain the full effect of frequency or reliability on mode choice, an alternative strategy to the use of waiting time as an explanatory variable could be to include frequency and reliability variables.

7.5.2 *Socio-economic data*

One of the major problems experienced with the data available for this study was the definition of a variable, or group of variables, that would explain moped availability. The number of mopeds per person aged 15 years or over did not perform well. Further work is necessary on this point, and to assist this additional age data will be required as well as, perhaps, direct information on the real abailability of a moped to each person. Adequate data to enable the determination of family or household life cycle are also necessary.

Personal income data should also be obtained as these could prove relevant to certain trip types, especially work and firm's business trips. The ownership of each car (private or company) as well as identification of the responsibility for paying car usage costs (company or not) should be determined.

It would also be of value to record whether the traveller using public transport is travelling at his employer's expense or whether it is a personal tax-deductable cost, since both of these situations occur in the Netherlands. A further item of relevant information is that of the types and ages of cars owned.

7.5.3 *Trip data*

For most purposes, certainly for destination- and mode-choice models, the trip data from the VODAR study are adequate. For shopping trips, however, additional information on the trip, such as the time spent at the shopping centre, number of shops visited, how much money was spent and the types of goods bought, would be relevant. For frequency–choice modelling, data for a period of at least one week should be obtained, so that a reasonably reliable estimate of trip frequency can be obtained.

One of the questions pertinent to disaggregate modelling is the identification of relevant alternatives. The philosophy adopted in this study was to aggregate across a group of people and to assume that the total range of choices made by that group represented the total range of relevant alternatives available to each individual. However, this procedure is relevant only if a large number of interviews is available from a particular area. If this is not the case, or if a more personal definition is considered preferable, then the interview respondent should be asked to identify all the alternatives available to him but rejected in favour of that chosen.

7.5.4 *Sample size*

The relatively small sample size necessary for the estimation of disaggregate models offers potential for important savings in both study costs and duration. Whilst the work undertaken in this study was relatively limited in nature, the indications are that for most models a sample of between 1,000 and 2,000 observations would be adequate. As a general rule, the greater the number of variables included in the model the larger the sample needs to be. One very important aspect of sampling is that the sample should be so selected as to ensure a high degree of variability within it.

The sample size, in terms of the number of households, could be decreased somewhat by increasing the period for which trip data are collected, thus effectively increasing the number of trips recorded per household. Whilst many regular trips, such as those to work and school, would be repeated, most of the less regular, or irregular, trips would not, and it is generally to obtain sufficient information on such trips that the full sample is required.

A further aspect of sampling is the problem (discussed above) of the identification of relevant alternatives. If a clustered sample is utilised, relevant alternatives can be identified by analysis of the total set of interviews, whereas if totally random samples are utilised, data on relevant alternatives must be obtained in the course of the interview or inferred from other information.

7.6 Possible extensions of research on disaggregate and simultaneous models

We consider that the results of this study, as well as of work undertaken by others, are such as to justify continued research and development on disaggregate simultaneous travel demand models. In so far as such work is directed towards the improvement of the conventional urban transportation model system (UTMS), it is our opinion that the primary concern should be with incremental improvements rather than sudden, radical changes. Independent, therefore, of the discussion which exists around the theory underlying Wilson's model, the similarity between its ultimate specification and the more traditional gravity and modal split models on the one hand, and the specific form of logit on the other, is one area that appears to justify further attention.

Another area to which effort should be devoted is that of transferability. If a disaggregate model is truly a behavioural model, and if it has been estimated with data in which a high degree of variability exists, then it can be expected that that model can be used in different geographic locations and for populations with different economic structures without amendment of the coefficients. If this is so, then many of the resources devoted to data collection in traditional urban transportation studies could be beneficially diverted to other activities. In view of data collection costs in relation to total study costs, empirical work on this area is therefore of no little importance.

Most of the research conducted to date has been concerned with work and shopping trips. These two purposes, however, represent less than half of the total daily urban trips. Many other trip purposes present greater problems in modelling at the disaggregate level, especially in the definition of alternative destinations. This is therefore another potentially fruitful area for research.

Last but not least of the areas which we consider to be of prime importance within the context of traditional study procedures is that of the transformation of a disaggregate model into an aggregate model for prediction purposes. Whilst the theoretical aspects of this problem should not be overlooked, the plea for an incremental approach to improving the conventional UTMS suggests that an empirical approach is also valid.

In the longer term, additional research can be done if better data are available, of the type described in this chapter, for instance.

8 Conclusions and Recommendations

8.1 General conclusions

The general, and most important, conclusion that can be drawn from the work undertaken in the course of this study is that disaggregate simultaneous models, estimated utilising the multinomial logit model, have considerable potential in the field of transportation, or travel demand, modelling. Not only do they answer in many respects the criticisms of traditional approaches quoted in Chapter 1, but they also emphasise many of the other very important technical deficiences inherent in those approaches. In addition to application in comprehensive planning studies, they are highly relevant to policy studies and both local and specific studies; indeed, the shopping model described in this report clearly could be developed into a full shopping model for use in shopping centre location studies. They offer major advantages over conventional aggregate modelling techniques, without, so far as is yet apparent, any significant disadvantages. There would therefore seem to be ample justification for extending the type of work described in this report to cover a wider range of problems relevant to both the Netherlands and other countries.

8.2 The policy implications of the models estimated

The policy implications which can be drawn from the models estimated, given the limitations of the data from which they were estimated, are of some importance. The most significant conclusions which can be drawn are:

1 Mode-choice for single home–work chains or shopping trips in a medium-size Dutch town appears not to be influenced by travel cost. This implies that increases in fuel or parking costs, or increases or decreases in public transport fares, will have little impact on modal split.
2 Terminal times – that is, the time taken in walking to or from a car, bus, train, etc. – are weighted more heavily than in-vehicle travel times.

The one exception is waiting and transfer time for bus and train, but the relative value of waiting time estimated in this study could be unreliable because of data deficiencies. What is particularly important is the high inconvenience of walking to or from a parking place (for bicycle, moped or car), which was estimated as being five times as inconvenient as in-vehicle travel time; this implies that the use of public transport can be positively influenced by making public transport more accessible, and by locating car parks away from the ultimate destination -- on the edge, rather than in the middle, of town centres, for example.

3 Car availability explains most of the choice of car as a mode, and becomes increasingly important with the length of the trip. This implies that one of the most effective ways of influencing mode choice is through car ownership.

4 The decision on whether to go home at lunch or not is, as might be expected, highly dependent upon distance. Furthermore, the choice of mode for work trips and the choice of whether or not to go home at lunchtime (one or two chains) are highly simultaneous, or interdependent. This implies, for example, that a reduction in the number of work–home–work trips at lunch-time would accompany a switch from car, moped or bicycle to public transport.

5 Level of service – i.e. the characteristics of the transportation systems – appears to be more important in the choice of destination and mode for shopping trips than it is in the choice of mode for work trips. This has implications similar to, but stronger than, point 2 above.

6 The relative attractiveness of alternative shopping centres is non-linear with size; relative attractiveness increases at a steadily decreasing rate as size increases.

One general conclusion which can be drawn from the estimated models is that no evidence has been found to suggest that marginal rates of substitution (the relative weights assigned to different variables by people making trip decisions) are necessarily the same for all types of trip; considerably more work is required, however, on different sets of data before any definite set of general conclusions on marginal rates of substitution can be drawn.

8.3 Recommendations for future work

In view of the general conclusion that disaggregate simultaneous travel demand models offer major advantages over traditional travel demand models, it would seem highly desirable to implement a continuing programme of research and development. This programme should be directed towards the introduction of disaggregate models as an element of the standard transportation study procedures, whereby they can become a common technique available to all transportation planning studies.

A series of possible topics for further work have been identified and discussed in section 7.6 of this report. These vary from relatively low cost studies, which could be undertaken and completed quickly, to major studies necessitating the collection of new data. The implementation of disaggregate and simultaneous models is, however, not dependent upon the completion of major new studies. Many improvements in traditional methods can be introduced relatively quickly.

Although disaggregate models appear to place very much greater demands on the qualities of the data required for both model estimation or calibration and prediction, these demands relate equally to aggregate techniques; but, because they are generally implicit rather than explicit, as is the case with disaggregate models, they are rarely recognised. The fact that they become obvious through the use of disaggregate techniques serves only to strengthen the argument that many of the data currently provided as input to transportation forecasts are of extremely dubious origin and value. If meaningful predictions of travel demand are to be prepared, then much more attention must be given to the forecasting of the socio-economic structure of communities.

Appendix

Summary of variable codes

CODE	*Description; alternatives to which applicable*
BFCON	Constant; moped
BFHHINC	Household income; moped
BFHHINC/PER	Household income divided by number of persons; moped
BFMOA	Moped ownership (number of mopeds divided by number of persons aged 15 years or more); moped
BFMOA1	Moped ownership; single chains, moped
BFMOA2	Moped ownership; double chains, moped
BFOCC	Occupation; moped
BFOVTT	Total out-of-vehicle travel time; moped
CAOD	Car availability (number of cars divided by number of licensed drivers); car
CAODEI	Car availability; shopping trips to Eindhoven, car
CAODLOK	Car availability; shopping trips to local centres, car
CHHINC	Household income; car
COCC	Occupation; car
COVTT	Total out-of-vehicle travel time; car
EICON	Constant; shopping trips to Eindhoven
EMP	Retail employment; shopping trips
EMPEI	Retail employment; shopping trips, Eindhoven
EMPLOK	Retail employment; shopping trips, local centres
FBOP	Bicycle ownership (number of bicycles divided by number of persons aged five years or more); bicycle
FBOP1	Bicycle ownership; single chains, bicycle
FBOP2	Bicycle ownership; double chains, bicycle
FCON	Constant; bicycle
FHHINC	Household income; bicycle
FHHINC/PER	Household income divided by number of persons; bicycle
FOCC	Occupation; bicycle
FOVTT	Total out-of-vehicle travel time; bicycle

HHINC	Household income
HHINC/PER	Household income divided by number of persons in household
IVTT	In-vehicle travel time; all modes except walk
IVTT1	In-vehicle travel time; single chains, all modes except walk
IVTT2	In-vehicle travel time; double chains, all modes except walk
LCON	Constant; double chains
LTT	Lunch-time travel time; double chains
OPTC	Out-of-pocket travel costs; car, moped, bus and train
OVTT	Total out-of-vehicle travel time
POVTT	Parking out-of-vehicle travel time; car, bicycle and moped
POVTT1	Parking out-of-vehicle travel time; single chains, car, bicycle and moped
POVTT2	Parking out-of-vehicle travel time; double chains, car, bicycle and moped
PTCON	Constant; bus and train
PTHHINC	Household income; bus and train
PTHHINC/PER	Household income divided by number of persons; bus and train
PTOCC	Occupation; bus and train
SOVTT	Waiting and transfer time; bus and train
TWOCC	Occupation; bicycle and moped
WOVTT	Walk time; walk
WSOVTT	Walking time to and from bus stop or station; bus and train

Bibliography

Becker, G.M., de Groot, M., and Marschat, J., 'Probabilities of choices among very similar objects: an experiment to decide between two models' *Behavioral Sciences* vol. 8, 1963.

Ben-Akiva, M.E., 'Structure of passenger travel demand models', PhD thesis, Department of Civil Engineering, MIT, Cambridge, Mass., 1973.

Ben-Akiva, M.E., and Lerman, S., 'Some estimation results of a simultaneous model of auto ownership and mode choice to work' *Transportation* vol. 3, no. 4, 1974.

Bouchard, R.J., *Relevance of Planning Techniques to Decision Making,* HRB Special Report no. 143, Highway Research Board, Washington DC 1974.

Brand, D., *'Theory and method in land use and travel forecasting' Highway Research Record* no. 422, Highway Research Board, Washington DC 1972.

Brand, D., *Travel Demand Forecasting: Some Foundations and a Review,* HRB Special Report no. 143, Highway Research Board, Washington DC 1973.

Bureau of Public Roads (BPR), *Calibrating and Testing a Gravity Model for any size Urban Area,* US Department of Commerce, Washington DC 1965.

Buro Goudappel en Coffeng, *VODAR, Huisenquête: Bewerking Gegevens,* prepared for the Rijswaterstaat, The Hague 1974.

Buro Goudappel en Coffeng, *VODAR, Huisenquête: Ontwerp en uitvoering,* prepared for the Rijkswaterstaat, The Hague 1971.

Buro Goudappel en Coffeng, 'NATS stage 1: the distribution and modal split model', prepared for Norwich City Council, Norwich 1972 (unpublished).

Cambridge Systematics, Series of unpublished technical memoranda on the automobile ownership project, Cambridge, Mass., 1973–74.

Centraal Bureau voor de Statistiek (CBS), *Systematische bedrijfsindeling SBI 1970,* The Hague 1970.

Centraal Bureau voor Statistiek, *Systematische beroepsindeling 14e Algemene Volkstelling*, The Hague 1971.

Centraal Bureau voor Statistiek, *Statistiek van het personenvervoer 1971,* Staatsuitgeverij, The Hague 1973.

Charles River Associates (CRA), *A Model of Urban Passenger Travel Demand in the San Francisco Metropolitan Area,* prepared for the California Division of Bay Toll Crossings, 1967.

Charles River Associates, *A Disaggregate Behavioral Model of Urban Travel Demand,* Federal Highway Administration, US Department of Transportation, Washington DC 1972.

Cheslow, M.D., 'The use of intercity multimodal forecasting models by the US Department of Transportation', paper prepared for presentation at the International Conference of Transportation Research, Bruges 1973.

Commissie Bevordering Openbaar Vervoer Westen des Lands (COVW), *De vervoerswijzen van de inwoners van het gebied van onderzoek,* COVW, The Hague 1970.

Domencich, T.A., Kraft., G., and Valette, J.P., 'Estimation of urban passenger travel behaviour: an economic demand model' *Highway Research Record* No. 238, Highway Research Board, Washington DC 1968.

de Donnea, F.X., *The Determinants of Transport Mode Choice in Dutch Cities,* Rotterdam University Press, Rotterdam 1971.

Fleet, C.R., and Robertson, S.R., 'Trip generation in the transportation planning process' *Highway Research Record* no. 240, Highway Research Board, Washington DC 1968.

Henderson, J.M., and Quandt, R.E., *Microeconomic Theory,* McGraw-Hill, New York 1958.

Highway Research Board (HRB), *Urban Travel Demand Forecasting,* HRB Special Report no. 143, Washington DC 1973.

Hillegas, T.J., 'Urban transportation planning – a question of emphasis' *Traffic Engineering* vol. 19, no. 7, 1969.

House of Commons, *Second Report from the Expenditure Committee, Urban Transport Planning* vol. 1 ('Report and Appendix'), Her Majesty's Stationery Office, London 1972.

Hupkes, G., 'Delphi opinion poll FUCHAN 1 – Evolution of change affecting the passenger car' *Transportation* vol. 3, no. 1, 1974.

Koppelman, F., PhD dissertation in preparation, Department of Civil Engineering, MIT, Cambridge, Mass.

Kraft, G., *Demand for Intercity Passenger Travel in the Washington–Boston Corridor* part V, Systems Analysis and Research Corporation, US Department of Commerce, Washington DC 1963.

Kraft, G., and Wohl, M., 'New directions for passenger demand analysis and forecasting' *Transportation Research* vol. 1, no. 3, 1967.

Luce, R.D., *Individual Choice Behaviour,* John Wiley, New York 1959.

Luce, R.D., and Suppes, P., 'Preference, utility and subjective probability' in Luce, R.D., Bush, R.R., and Galanter, E. (eds), *Handbook of Mathematical Psychology* vol. 3, John Wiley, New York 1965.
McCarthy, G.M., 'Multiple regression analysis of household trip generation – a critique' *Highway Research Record* no. 297, Highway Research Board, Washington DC 1969.
McFadden, D., *The Revealed Preferences of a Government Bureaucracy,* Technical Report no. 17, Institute of International Studies, University of California, Berkeley 1968.
McGillivray, R.G., 'Binary choice of transport mode in the San Francisco Bay area', PhD thesis, Department of Economics, University of California, Berkeley 1969.
McGillivray, R.G., 'Mode split and the value of travel time' *Transportation Research* vol. 6, no. 4, 1972.
MacKinder, I.H., Rafferty, J., Singer, E.H.E., and Wagon, D.J., *Compact-2: A Simplified Program for Transportation Planning Studies,* MAU Note 201, Department of the Environment, London 1970.
McLynn, J.M., and Woronka, T., *Passenger Demand and Modal Split Models,* Northeast Corridor Transportation Project, US Department of Transportation, Washington DC 1969.
McIntosh, P.T., and Quarmby, D.A., *Generalised Costs and the Estimation of Movement Costs and Benefits in Transport Planning,* MAU Note 179, Department of the Environment, London 1970.
Manheim, M.L., 'Search and choice in transport systems analysis' *Highway Research Record* no. 293, Highway Research Board, Washington DC 1969.
Manheim, M.L., *Fundamental Properties of Systems of Demand Models,* Discussion Paper T 70-1, Department of Civil Engineering, MIT, Cambridge, Mass., 1970.
Manheim, M.L., 'Practical implications of some fundamental properties of travel demand models' *Highway Research Record* no. 422, Highway Research Board, Washington DC 1972.
Manski, C., 'Qualitative Choice Analysis', unpublished PhD thesis, Department of Economics, MIT, Cambridge, Mass., 1973.
Marschak, J., 'Binary-choice constraints and random utility indicators' in Arrow, K.J., *et al.* (eds), *Mathematical Methods in the Social Sciences,* Stanford University Press, Stanford 1959.
Nederlands Economisch Instituut (NEI), *Integrale verkeers- en vervoersstudie,* Staatsuitgeverij, The Hague 1972.
de Neufville, R., and Stafford, J.H., *Systems Analysis for Engineers and Managers,* McGraw-Hill, New York 1971.

O'Farrell, P.N., and Markham, J., 'Commuter perception of public transport work journeys' *Environment and Planning* vol. 6, no. 1, 1974.

Oi, W.Y., and Shuldiner, P.W., *An Analysis of Urban Travel Demands,* Northwestern University Press, Evanston, Ill., 1962.

Peat, Marwick, Mitchell and Co. (PMM), *Implementation of the N-dimensional Logit Model,* Comprehensive Planning Organization, San Diego County, California, 1972.

Plourde, R.P., *Consumer Preference and the Abstract Mode Model: Boston Metropolitan Area,* Research Report R68-51, Department of Civil Engineering, MIT, Cambridge, Mass., 1968.

Plourde, R.P., 'Development of a behavioral model of travel mode choice', PhD thesis, Department of Civil Engineering, MIT, Cambridge, Mass., 1971.

Pratt, R.H., and Deen, T.B., 'Estimation of a sub-modal split within the transit mode' *Highway Research Record* no. 205, Highway Research Board, Washington DC 1967.

Quandt, R.E. (ed.), *The Demand for Travel: Theory and Measurement,* Heath Lexington Books, Lexington, Mass., 1970.

Quandt, R.E., and Baumol, W.J., 'Abstract mode model: theory and measurement' *Journal of Regional Science* vol. 6, no. 2, 1966.

Quarmby, D.A., 'Choice of travel mode for the journey to work: some findings' *Journal of Transport Economics and Policy* vol. 1, no. 3, 1967.

Reichman, S., and Stopher, P.R., 'Disaggregate stochastic models of travel mode choice' *Highway Research Record* no. 369, Highway Research Board, Washington DC 1971.

Richards, M.G., Verberne, B.E.N., and Willemse, G.L.S., 'VODAR: some aspects of the home interview survey' *Traffic Engineering and Control,* May 1972.

Robinson, W.S., 'Ecological correlations and the behavior of individuals' *American Sociological Review* vol. 15, 1950.

Stopher, P.R., 'A probability model of travel mode choice for the work journey' *Highway Research Record* no. 283, Highway Research Board, Washington DC 1969.

Stopher, P.R., and Lavender, J.O., 'Disaggregate, behavioral travel demand models: empirical tests of three hypotheses' *Transportation Research Forum Proceedings 1972*

Stopher, P.R., and Lisco, T.E., 'Modelling travel demand: a disaggregate behavioral approach, issues and applications' *Transportation Research Forum Proceedings 1970.*

Stopher, P.R., Spear, P.D., and Sucher, P.O., 'Towards the development of measures of convenience for travel modes', paper prepared for presentation at the Annual Highway Research Board Meeting, Washington DC 1974.

Stowers, J.R., and Kanwit, E.L., 'The use of behavioral surveys in forecasting transportation requirements' *Highway Research Record* no. 106, Highway Research Board, Washington DC 1966.

Theil, H., *Principles of Econometrics,* John Wiley, New York 1971.

Warner, S.L., *Stochastic Choice of Mode in Urban Travel: a Study in Binary Choice,* Northwestern University Press, Evanston, Ill., 1962.

Watson, P., *The Value of Time: Behavioral Models of Modal Choice,* Heath Lexington Books, Lexington, Mass., 1974.

Weiner, E., 'Modal split revisited' *Traffic Quarterly,* January 1969.

Westin, R.B., 'Predictions from binary choice models' *Journal of Econometrics* vol. 2, no. 1, 1974.

Wickstrom, G.V., 'Are large-scale home-interview origin-destination surveys still desirable?', paper presented at the Transportation Forecasting Committee meeting at the Annual Highway Research Board Meeting, Washington DC 1971.

Wilson, A.G., 'Entropy maximizing models in the theory of trip distribution' *Journal of Transport Economics and Policy* vol. 3, no. 1, 1969.

Wilson, A.G., Hawkins, A.F., Hill, J.G., and Wagon, D.J., 'Calibrating and testing the SELNEC transport model' *Regional Studies* vol. 3, no. 3, 1969.

Wootton, H.J., and Pick, G.W., 'Trips generated by households' *Journal of Transport Economics and Policy* vol. 1, no. 2, 1967.

Index

The Authors

Martin G. Richards is Director with special responsibility for all quantitative study work in the field of transportation planning at Buro Goudappel & Coffeng B.V., Diventer, the Netherlands. He received his M.Sc. in Highway and Traffic Engineering at the University of Birmingham.

Moshe E. Ben-Akiva is with Cambridge Systematics Inc., Massachusetts, and is also an Associate Professor at the Massachusetts Institute of Technology.